Car Handyman

Steering & Suspension

Marshall Cavendish

CONTENTS

Editor Fred Milson
Designer Chris Walker
Production Controller Steve Roberts
Typesetting ABM Typographics Limited, Hull

Produced by Marshall Cavendish Books Limited
58 Old Compton Street
London W1V 5PA

ISBN 0 86307 355 7

Printed and bound in Milan, Italy by New Interlitho

INTRODUCTION

Most DIY mechanics tend to shy away from tackling repairs to either steering or suspension, imagining that it's far too complicated or dangerous. *Steering & Suspension* unravels some of the mysteries and shows you how to deal with these components simply and safely. Every aspect of the job is covered, from renewing shock absorbers to checking and replacing bearings and steering joints

Steering systems explained

Modern steering systems have to cope with a range of different stresses, from cornering at high speed to parking. Two main systems are used — how do they work?

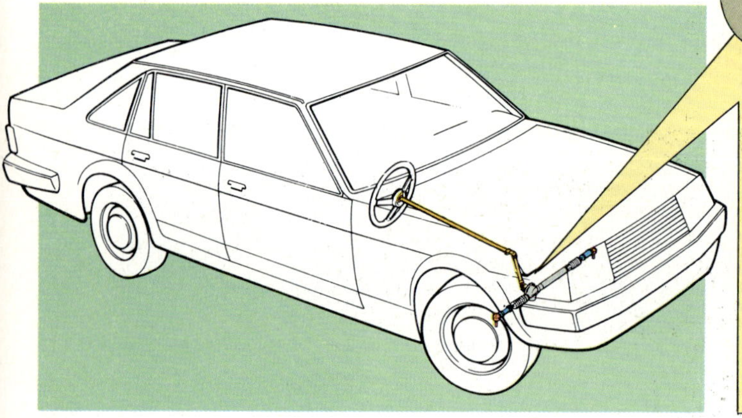

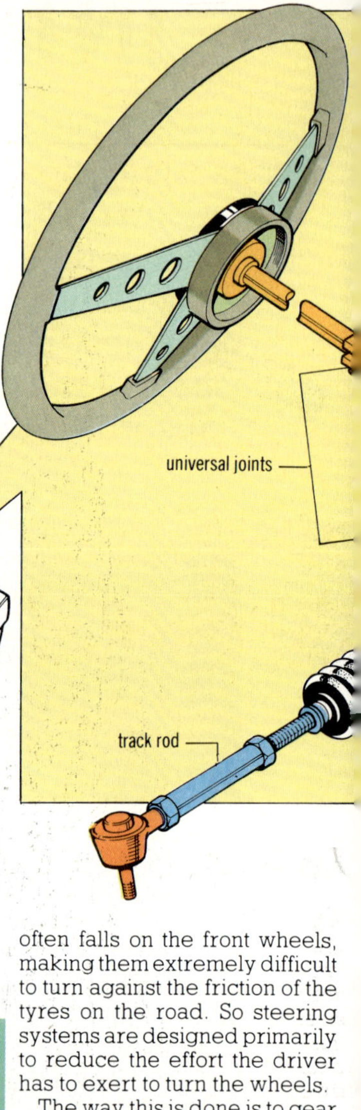

When all vehicles were horse drawn, steering was simply a matter of pointing the horse in the right direction. With a heavy, self-propelled vehicle like a modern car, things are unfortunately not that easy. All of the steering effort has to come from the driver operating a steering wheel linked to the front wheels.

At least half of the car's weight often falls on the front wheels, making them extremely difficult to turn against the friction of the tyres on the road. So steering systems are designed primarily to reduce the effort the driver has to exert to turn the wheels.

The way this is done is to gear the steering so that the movement of the steering wheel is not linked directly to the movement of the wheels. This means that it takes a number of turns of the steering wheel to move the road wheels a relatively small distance. The gearing also means that the driver can make slight changes in direction that would be almost impossible, even dangerous, with direct steering.

Q On some small cars the steering is really heavy, while on some larger vehicles it is light. Why is this?

A The way that the steering system is geared can reduce the effort needed to steer the car. But only to a limited extent: the more weight that is over the front of the car, the heavier the steering is likely to be. On some small cars — particularly front wheel drive models — the only place where the engine and gearbox can be positioned is in line with or slightly forward of the wheels. The result is that the steering is heavy. On larger cars there is usually room to spread the weight more evenly, with the result that the steering is often much lighter than you would normally expect it to be.

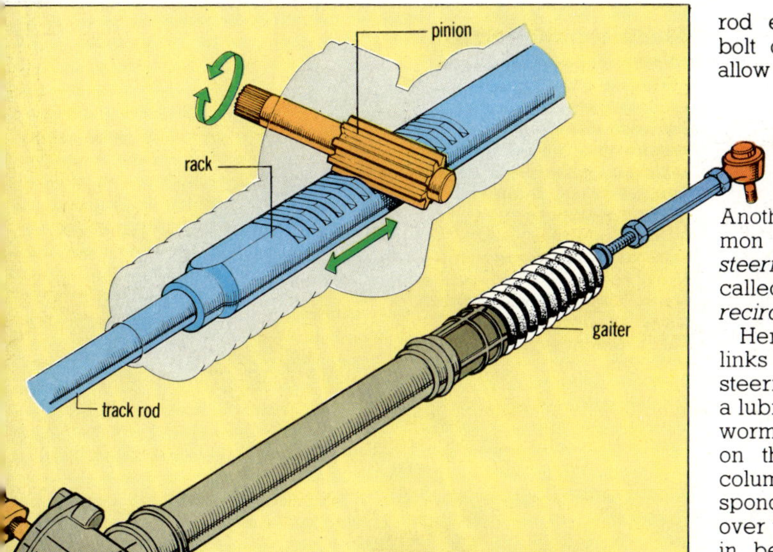

RACK AND PINION STEERING

Common to most European-built cars, this is a simple direct system with few moving parts. A small cog or pinion attached to the end of the steering column engages with a number of teeth on the steering rack

rod ends are ball joints that bolt onto the steering arm to allow for suspension movement.

Steering box system

Another steering system, common on Japanese cars, is the *steering box* system — often called the *worm and nut* or *recirculating ball* system.

Here, the mechanism which links the steering column to the steering gear is all contained in a lubricated and sealed box. A worm gear groove is machined on the end of the steering column. A nut with a corresponding internal groove slides over the top with ball bearings in between. As the steering shaft is turned, the nut moves up and down the shaft.

Connecting the steering box with the front wheels is a series of ball-jointed arms and track rods called the *steering linkage*. Operated by a drop arm from the output shaft of the steering box, the linkage exerts pressure on the steering arms at each axle.

Rack and pinion system

Most modern cars — particularly those made in Europe — are fitted with *rack and pinion* steering. Manufacturers favour this system because it is compact, with few mechanical parts; drivers like it because there is very little free play, which gives the steering a direct and precise feel.

In the rack and pinion system a small gear — the pinion — connected to the base of the steering column meshes with a toothed rack; both components are encased in a steel tube filled with lubricant. The rack is mounted across the car and is firmly bolted to the front cross members or bulkhead by two U-bolts or mounting brackets.

At each end of the rack housing is a *track rod* which carries the steering from the rack to the wheels. The track rods are usually encased in flexible rubber gaiters containing gearbox oil. The track

Steering feel

Both the rack and pinion and the steering box system are designed to pass a certain amount of road information on to the

Q I can move the steering wheel on my car quite a long way before the road wheels begin to move. Is this normal?

A If your car is fitted with the steering box system, a certain amount of play is almost unavoidable because of the ball-jointed links. But this should never be more than 2 in. (50 mm). The limit for cars fitted with rack and pinion steering is 1 in. (25 mm).

If the free play is greater than this, check the track rod ends and the ball joints — they may be worn and need replacing. Alternatively, any excessive play you feel may be because the universal joints on the steering column are worn — so check them and replace if necessary.

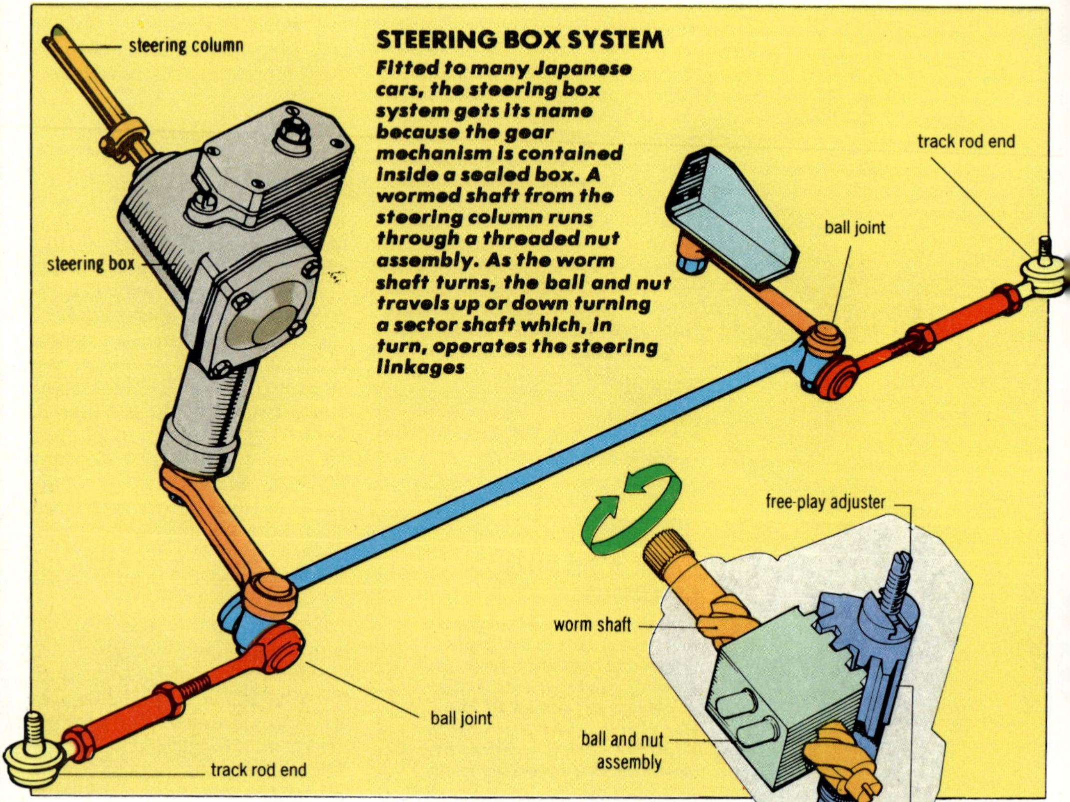

STEERING BOX SYSTEM
Fitted to many Japanese cars, the steering box system gets its name because the gear mechanism is contained inside a sealed box. A wormed shaft from the steering column runs through a threaded nut assembly. As the worm shaft turns, the ball and nut travels up or down turning a sector shaft which, in turn, operates the steering linkages

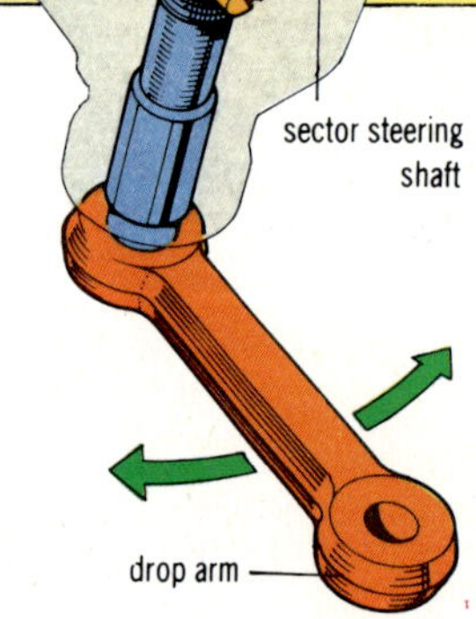

driver. For one thing, you can feel the progressive build-up of effort required as the wheel is turned farther and farther towards full lock — an essential safety feature when cornering at speed.

You can also clearly feel how the tyres are behaving. Any loss of traction or grip is immediately accompanied by a sudden lightening of the steering effort. If the car starts to skate or *aquaplane* on a sheet of water, for example, you will immediately feel that the wheels no longer grip the road.

Q Why do some cars have smaller turning circles than others?

A A small wheelbase can have an important bearing on the turning ability of a car — but smaller cars do not always have tighter turning circles. Design factors like restrictions on the space under the wheel arches, or problems with clearances between the wheels and the brake, suspension or transmission gear are just as important. They often prevent manufacturers getting as tight a turning circle as they would like. And it is often on the smaller models that design problems such as these occur — with the result that the turning circle has to be restricted.

Steering wheel

The steering wheel is a key component in any steering system. In only a very few cars it is possible to connect the column directly to the steering gear because they do not align directly. To get over this problem, one or two universal joints are introduced into the column. The joints are usually the Hooke-type, with two U-shaped yokes bolted to a cross-shaped inner piece, often made of rubber.

Power-assisted steering

Even though the steering is geared to make it easier to turn the road wheels, this isn't always enough — particularly on large front wheel drive cars which have a substantial amount of weight on the front wheels. So to reduce the steering effort, many manufacturers fit power steering, where the force which the driver applies to the wheel is supplemented by power from a hydraulic ram.

Power assistance can be applied to rack and pinion or steering box steering, depending on whether the designer is going for the precision of rack and pinion or the freedom from kickback with a steering box.

The hydraulic pressure is supplied by a belt-driven pump fitted to the front of the engine. Once you turn the steering, valves open in the steering and the pressure is applied.

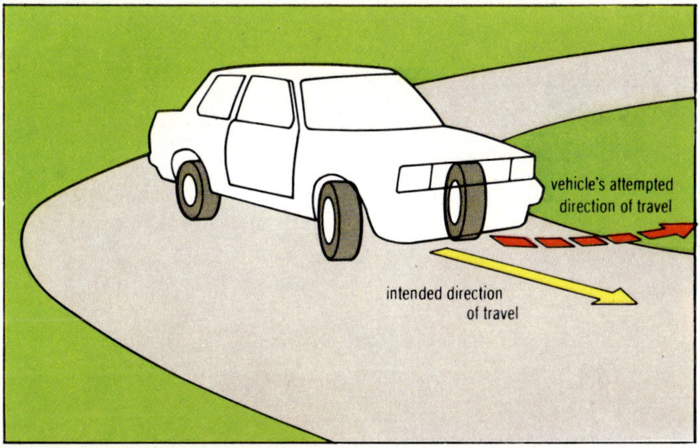

Opposite steering lock applied to correct oversteer

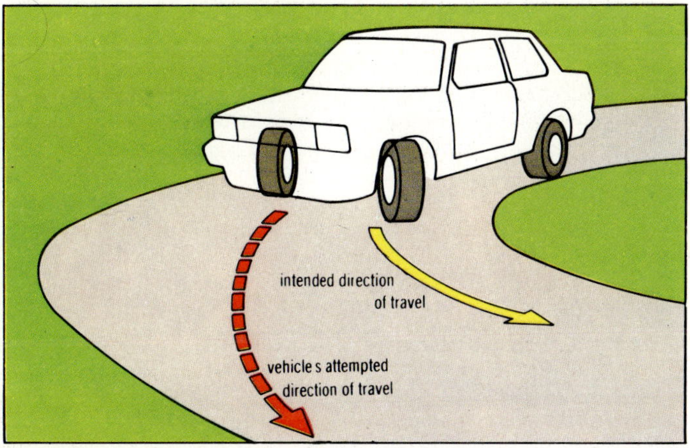

Extra steering lock applied to correct understeer

Q What does the term 'kickback' mean?

A Road wheels are often subjected to massive shock forces — for example, when the car runs over a large pothole or a tyre blows out at speed. The resulting jolt you feel at the steering wheel is known as kickback.

The amount of kickback you get depends on how much road feel you get through the steering system.

Q What is meant by the terms 'oversteer' and 'understeer'?

A Oversteer and understeer are usually only felt when cornering at high speed and occur when the front and rear wheels do not grip the road to exactly the same degree.

A car oversteers when it tries to spin into a corner, its rear wheels sliding out sideways. A car understeers when it tries to carry on in a straight line despite the driver's attempts to steer it into a corner.

Although oversteer and understeer both affect car control they are not specifically linked to the steering system but relate much more closely to the way a car is loaded, suspension geometry and to whether it is front or rear wheel drive.

For safety, most new cars are designed with understeer.

Checking and replacing dampers

Car's handling not as good as it used to be? Does it dip and wallow on corners and under braking? Try checking your dampers

Dampers, also known as shock absorbers, are a vital but often ignored part of the car's suspension system. Because they are hidden from view and go without maintenance for a number of years, dampers tend to be forgotten completely. Dampers wear out slowly over a long period of time and because of this it is very difficult to notice any change in their performance, but if they are worn you will certainly be able to tell the difference when you fit a new set. The car will be much more stable in cornering and will give a better ride.

When to do this job
When the car's ride and handling becomes noticeably bouncy
If the suspension makes strange noises

What this job involves
Checking the dampers visually
Carrying out damper tests
Removing and replacing telescopic and lever arm dampers

Points to watch
Dampers are now checked during the MoT test
Some garages now have an electronic machine to test dampers

To do this job
Tools: Spanners or sockets to fit damper mounting nuts or bolts; soft mallet or hammer and block of wood; jack; ramps or large blocks of wood; screwdriver (maybe); coil spring compressors (maybe)
Materials: Penetrating oil; new dampers (maybe); new rubber bushes (maybe); damper fluid (maybe)
Time: 10 minutes to do the bounce test — half a day to replace all four dampers
Degree of difficulty: Difficulty varies from car to car but is generally awkward rather than difficult

If you have the job professionally done . . .
Does the car handle and ride better? Do the dampers look clean and new?

STEP 1 TEST THE DAMPERS IN PLACE

A good test for worn dampers is to take the car for a ride down a bumpy road with sharp bends. Do not have the car too heavily loaded or it will be difficult to tell if the dampers are worn or not.

When you go round a corner, the car should not tilt to one side, or 'roll', excessively. Nor should it feel as if it is floating, or 'wallowing' when driven over bumps. If either rolling or wallowing occurs, the steering will also tend to feel vague at the same time and both are indications that the suspension is not doing its job.

If driving over a large bump or pothole causes the car to crash down on to its bump stops with a resounding thump and a 'graunching' noise you can be sure that the dampers are worn and require attention.

If a road test is not possible, or is not conclusive, lean on the front wing and push it down as far as you can. You should be able to feel the dampers firmly resisting your efforts. Faulty dampers will either allow you to bounce the car or resist all movement completely. For comparison, try this test on a newer car of the same model.

> **TIP**
>
> **The Bounce Test**
>
> **A quick and easy damper test is to rock the car violently up and down and then let go. The car should bounce up and then down just once or twice before it settles to rest. Do the test at each corner — if the car bounces up and down repeatedly the damper at that wheel needs changing.**

Even if the dampers seem to be performing properly it is well worth carrying out a further visual check for worn bushes and fluid leaks, as described in Step 3. This will enable you to spot a damper that is about to fail and which can be repaired before the car's handling begins to suffer.

Dampers should, ideally, always be replaced in pairs — one on each side of the car. Unbalanced dampers will affect a car's handling as adversely as a damper which is worn out. But if your car is relatively new, or has only recently had its dampers replaced, fitting a single new unit may not cause any problems. If in doubt, try fitting a single replacement and then test drive the car very carefully. You may find that the new unit simply highlights the poor performance of the remaining old damper and results in handling as imprecise as that previously caused by the failed unit. If so, fit another new unit as soon as possible.

1. Carrying out the bounce test with the dampers on the car

STEP 2 FIND THE DAMPERS

To find and identify the dampers used on your car, first check your car's manual. But you should find that with the car jacked up and wheel removed they are easy to spot. Even if you want to work on more than one damper it is best to tackle them one at a time. Not only does this make the job easier but it leaves you a handy reference when you come to reassembly. So jack up the car at the appropriate corner and support it safely on axle stands placed under the chassis.

Take off the road wheel and

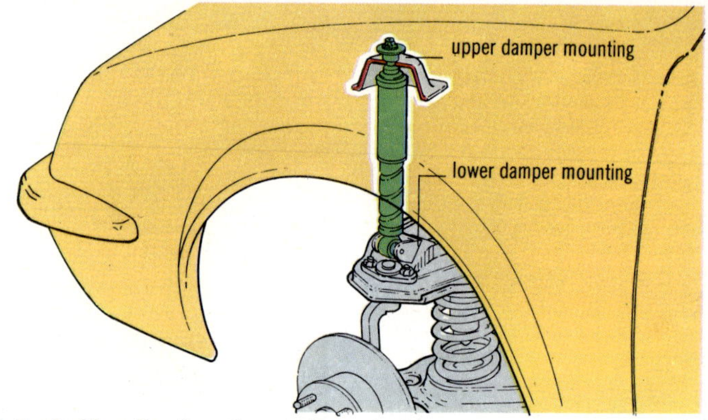

1. Typical location for a front telescopic damper

STEP 3 CHECK THE DAMPERS

To check for wear, hold the damper firmly near its mounting and try to move it. The forces acting on the suspension are huge so pull the damper hard. But do not confuse worn mounting bushes with wear in the damper itself. If there is any free play at either end between the two parts of the damper it must be renewed as soon as possible. This is very important with lever arm dampers which form part of the suspension linkage because free play in the damper will impair steering.

If you can move the damper as a whole away from the body or suspension then the rubber mounting bushes are worn or the mounting bolts are loose. If the mounting bolts are tight and the damper is in otherwise good condition you may find that you can renew the rubber bushes — so find out if your local accessory shop will sell you the bushes separately.

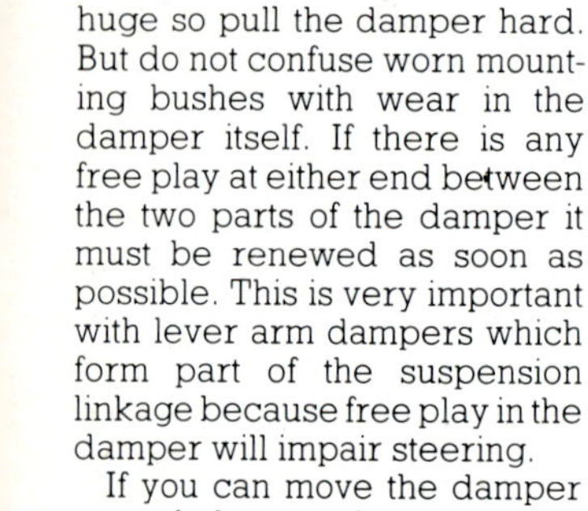

1. Testing bush for play

While you are looking at the damper mounts also check the suspension links and bodywork to which the dampers are attached . If you find any serious rusting here you should have the car checked by a garage as new mounts may need to be welded into place.

Finally, inspect the damper closely for any signs of a leak. If a telescopic damper is leaking it must be replaced, but in the case of lever arm dampers first make sure that the leak is not coming from the filler plug.

If the filler plug is leaking renew the washer and top up the fluid level in the dampers (**fig 5**). To do this you will need a suitable damper fluid which you can get from your dealer or accessory shop, though it is becoming a little hard to find.

The dampers are filled through a hole near the top of the body of the unit. This hole is sealed by a hexagon headed plug which you have to remove with a spanner.

You can top up most lever arm dampers without removing them from the car but with some types, such as the rear units on

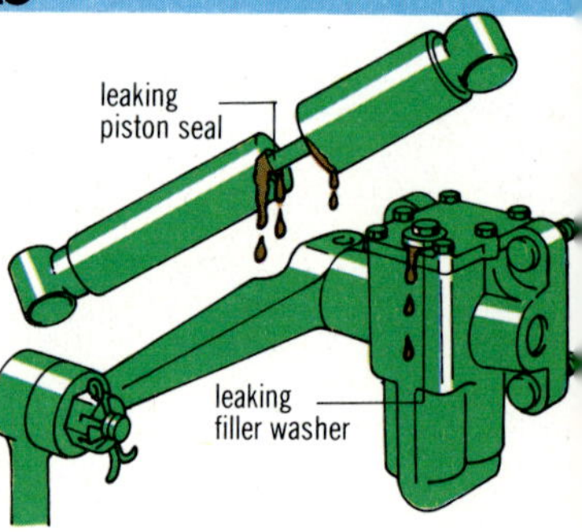

3. Leaking dampers — both types

the Morris Minor, you have to take them off the car to do the job (see Step 6).

Before you start refilling, ensure that the damper unit and filler plug are absolutely clean and that no dirt can enter the damper.

Take out the filler plug and pour the fluid into the damper until the level is at the bottom of the threaded hole or, if a level plug is fitted, until fluid is level with the hole.

2. Perished rubber bush

look into the wheel arch. You should see the damper mounted close to the wheel hub and fitted between the car's body and suspension.

Some older BL cars have the lever arm damper forming the top link of the front suspension, while telescopic dampers will be inside or near the coil springs, if fitted. To get a good look at the rear dampers, either telescopic or lever arm type, you may find it easier to work from underneath the car so make absolutely sure it is safely supported first.

2. Lever arm damper in place at the rear of the car

The fluid should come in a plastic bottle with a tube attached which makes it easy to fill the damper, but if not use a length of rubber hose attached to a funnel. As you fill the damper pump the lever arm to get any air bubbles out of the fluid. If the damper is still on the car get someone to push gently up and down on the wing — but do not do this with someone under the car.

If you have to unbolt the damper, you will find it easy to fill the unit with one hand and pump it with the other.

When the unit is full, fit the plug tightly and bolt the dampers back to the car if you removed them (see Step 6).

Now repeat the bounce test. If the dampers do not perform properly or if they still leak they will have to be renewed.

5. Topping up lever damper

4. Rusted mounting behind lever arm damper

TIP

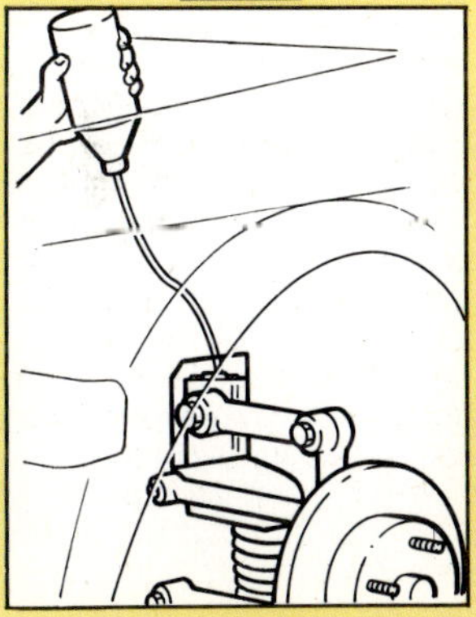

Squeeze it

If you find that it is difficult to get to the filler hole on a lever arm damper, put the new fluid in to a clean plastic bottle. Fit a length of plastic or rubber tube to the nozzle and you will be able to pump in the fluid easily.

TELESCOPIC DAMPER REMOVAL

1. With the wheel off, the suspension drops

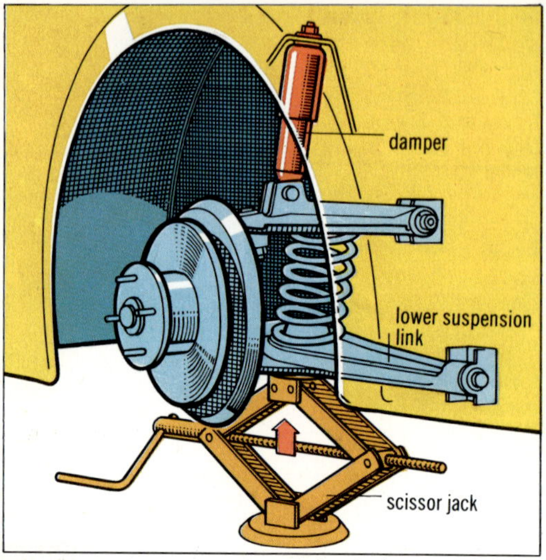

2. Compressing suspension with a jack

Telescopic dampers may be mounted within a coil spring or separately. If the spring and damper look as though they are one unit, remove them together and then separate them as described in Alternative 5.

Begin the job by removing the large nut and bolt holding the damper at each end. Take note of any washers or spacers and remember their positions for reassembly. Then take off the damper using a pulling and twisting action to ease it off. If the damper is particularly difficult to remove you may find it necessary to take the load off the suspension by using a jack to raise the lower suspension link by a small amount (**fig 2**).

The rubber bushes can also prove rather stubborn, and you may find that the metal inner bushes have corroded on to the bolt or stud. If the bush pulls away leaving the metal sleeve behind, soak the sleeve with penetrating oil and then try to twist it off using Mole grips.

You will probably find a large washer behind the damper on the other side of the rubber bush. It is essential that you replace this correctly and that you do not use one of a smaller

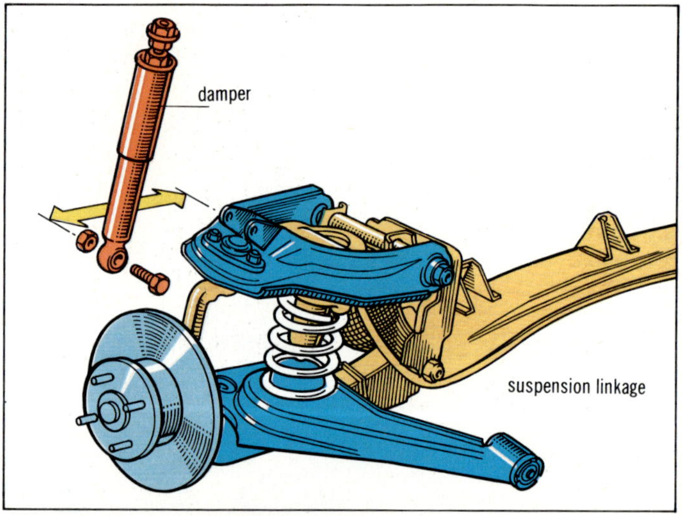

3. One type of damper lower mounting

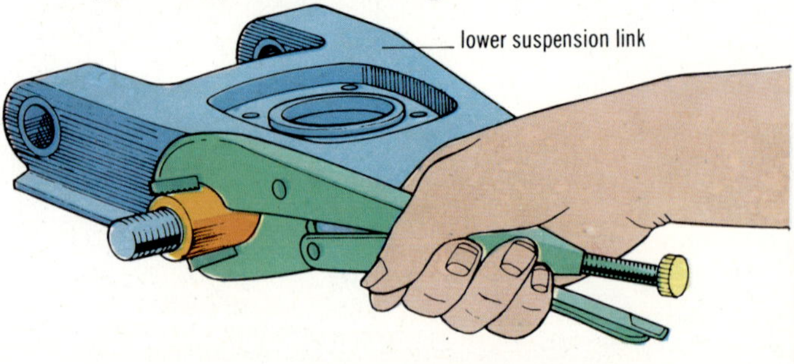

4. If a sleeve sticks to the mounting, remove it with grips

size, as the washer stops the 'eye' of the damper from slipping over the rubber bush.

Telescopic dampers, which are fitted to the majority of family cars, are not designed to be serviced or overhauled. When they are worn or damaged you must replace them with new units. It is, however, worth shopping around as replacement dampers are often heavily discounted. But buy a well known brand and beware of reconditioned units.

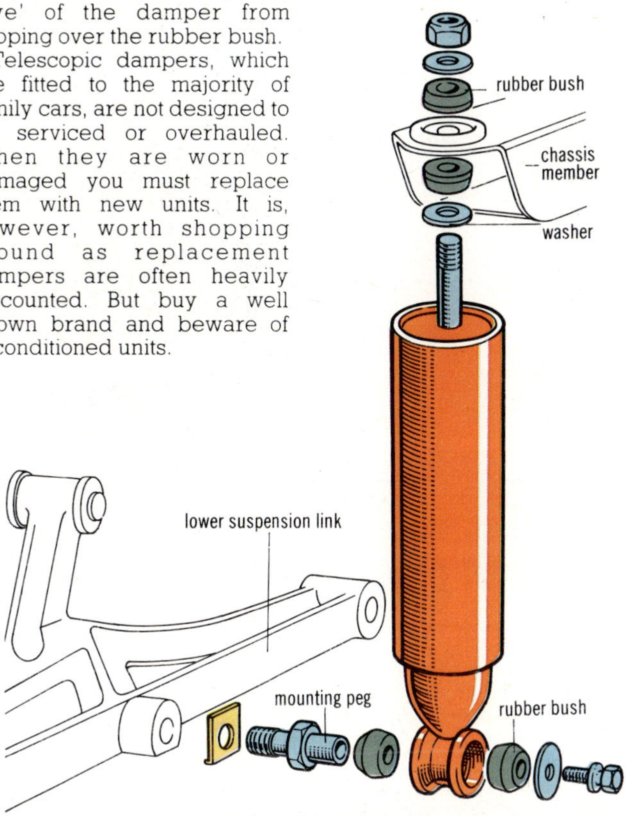

6. Bush and washer arrangements on damper mountings

5. Combined spring and damper

STEP 5 REPLACE THE TELESCOPIC DAMPERS

Replacing a telescopic damper is basically a reversal of the removal procedure. But before fitting a new damper it is worth operating the unit a few times by hand to check that it is working properly. To do this compress and extend the damper by hand. The damper should work smoothly and stiffly when operated slowly — but it should resist any attempt to compress or extend it rapidly.

Having established that the replacement damper is operating correctly, apply a little soap to the bushes to help them slide on, then fit it on to its mounting studs or bolts. Make sure that you fit all the washers and spacers in the same order that they were removed. If you are unsure of the way the unit fits back on, simply take a look at the damper on the other side of the car.

HOW IT WORKS

Shock absorber

Most modern cars are fitted with telescopic dampers which consist of two tubes which contain oil. When the car goes over a bump the oil is forced slowly from one tube to the other through small holes. This means that the damper cannot be compressed or extended rapidly and counteracts the tendency for the car to bounce on its springs.

Lever arm dampers work in the same way as telescopic ones but instead of comprising two oil-filled tubes, the damping fluid is contained in a chassis-mounted square-shaped reservoir which has an arm that pivots up and down, forming the top link of the front suspension.

ALTERNATIVE 5 TAKE OFF THE SPRING

On some cars the suspension coil spring is mounted on the damper body and the whole unit comes off together. However, care must be taken when you dismantle this type because the spring is under tension and must be released gently.

Unfortunately you will have to use proper spring compressors to get the spring off — you can hire these quite cheaply from tool hire and larger accessory shops. You may be able to compress and remove the spring by hand but this is tricky, dangerous and needs at least two people — furthermore, replacing the spring is harder.

Put the compressors on to the spring (you need at least two), hooking them over the top and bottom coils. The idea is to try and compress as many coils as possible (**fig 2**).

Screw the compressors up until the spring is pulled away from the top of the damper unit. You should now be able to see how the top of the spring is held down. If there is a split collet, flick this out with a screwdriver and save the pieces for later. If there is a nut, unscrew it. With other types the entire cap assembly will unscrew.

Take off the cap (**fig 3**) and its washers, making sure that you keep them in the right order. You can now carefully slacken off the compressors and release the tension on the spring. With the spring removed you can now treat the damper as a normal telescopic one.

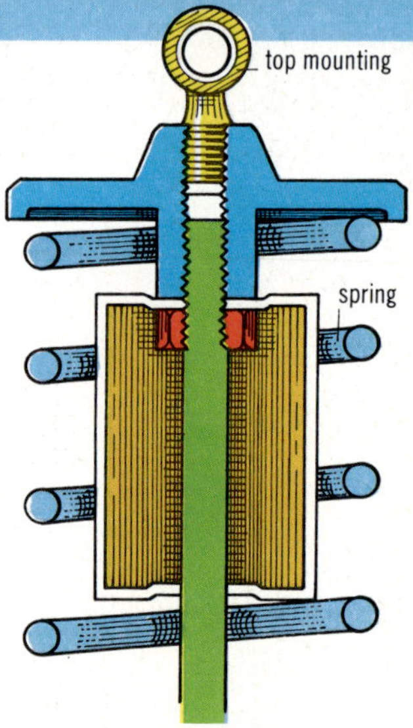

1. Section through top mounting

STEP 6 LEVER ARM DAMPER REPLACEMENT

Because lever arm dampers sometimes form part of the load-bearing suspension linkage, more care must be taken when removing them. Whether you are taking them from the front or rear of the car, the suspension must not be at the limit of its travel when you take them off.

With the car's chassis supported on axle stands or ramps, load the suspension slightly at the back by jacking up the rear axle an inch or so, or at the front by lifting the lower link of the suspension in the same way by about the same amount. But take care to load the suspension slowly or you could jack the car off its stands.

Next, disconnect the lever arm. The nut which holds it to the suspension will probably have a split-pin through it. If so, straighten this out and pull it from the nut with a pair of pliers. If the arm is difficult to remove from the suspension, try jacking the lower link up and down while you keep pulling. If it is stuck, you can jar the arm with a soft mallet or use a hammer and block of wood. You may find that the arm will rise up to its fullest extent when you have disconnected it. This is due to fluid pressure inside the unit and is quite normal.

With the arm disconnected, undo the bolts holding the damper body to the car (**fig 2**) and lift it away. It is best to

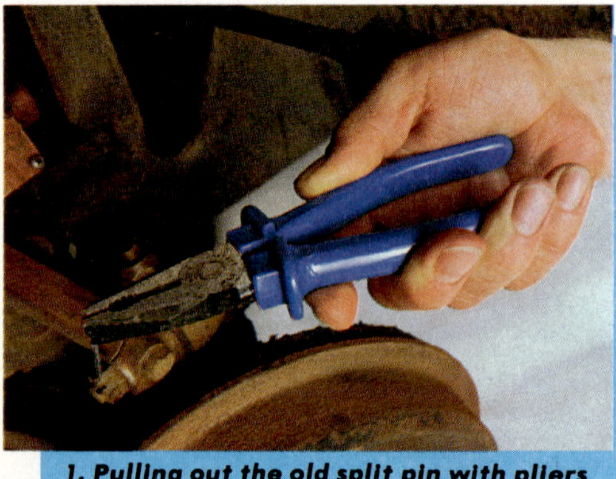

1. Pulling out the old split pin with pliers

2. Undoing damper mounting bolt

2. Tightening coil spring compressors

3. Lifting off the spring retaining cap

leave the lower suspension arm supported on the jack while you do the rest of the job. This will prevent the suspension upright falling outwards and straining the brake hose.

When rear lever arm dampers are fitted the dampers are held to the underside of the car by three or four bolts. The lever arm is bolted to a link which is attached to the rear axle. Undo either end of this link — whichever you find easier — then unbolt the unit from the car and lift it away.

As with telescopic dampers, it is worth shopping around for replacements for the lever arm types but note that you can get reconditioned lever arm dampers quite cheaply.

When fitting the new unit, bolt the damper to the car's body first, then make sure that the lever arm is working smoothly. Attach the end of the arm to the suspension but make sure that you use a new split-pin on the nut when you fit it. When everything is tight, lower the car to the ground and test drive the car.

3. Supporting suspension before removing damper

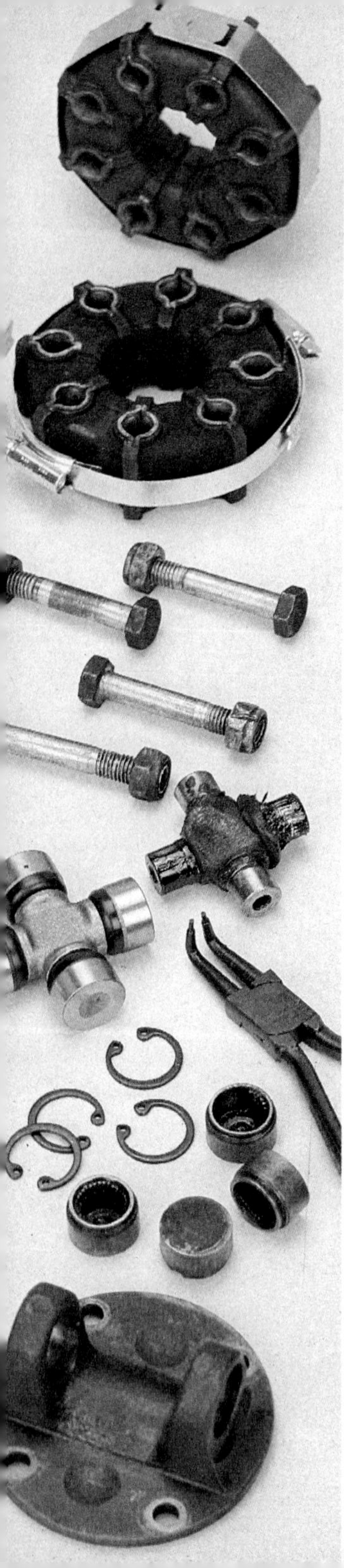

Cures for drive-shaft shudder

A knocking noise from the transmission may be caused by worn universal joints which you should renew or overhaul

A knocking or juddering sensation felt when accelerating or decelerating is usually a sign that a universal joint in the transmission requires attention. In most cases a quick inspection will confirm the diagnosis and appropropriate action can then be taken.

Replacing a universal joint is generally a straightforward job. But even if you decide not to tackle the job yourself and take your car to a professional, you may still make a small saving by pointing the mechanic straight to the cause of the problem so that he can cure it more quickly.

When to do this job
If there is knocking or juddering coming from the transmission
When a universal joint is worn out

What this job involves
Checking universal joints for wear
Removing worn joints
Overhauling and replacing universal joints

Points to watch
Special kits are now available to help you overhaul a propshaft that is not fitted with circlips. But these take quite a lot of skill. Do not buy one unless you are pretty confident. Some front and rear driveshafts are fitted with Hooke's joints that can be serviced in just the same way as a Hooke's in a propellor shaft
Fit new self-locking nuts every time that you bolt the prop shaft flanges back together

To do the job
Tools: Socket set; screwdrivers; hammer; workbench with vice; circlip pliers or punch
Materials: New universal joint; grease such as Castrol LM; new nuts for propshaft flanges
Time: Allow up to two hours for each universal joint
Degree of difficulty: Easy on most cars

If you have the job professionally done ...
Has the juddering or clonking disappeared?
Does the car vibrate at high speed?

STEP 1 FIND THE SOURCE OF THE JUDDER

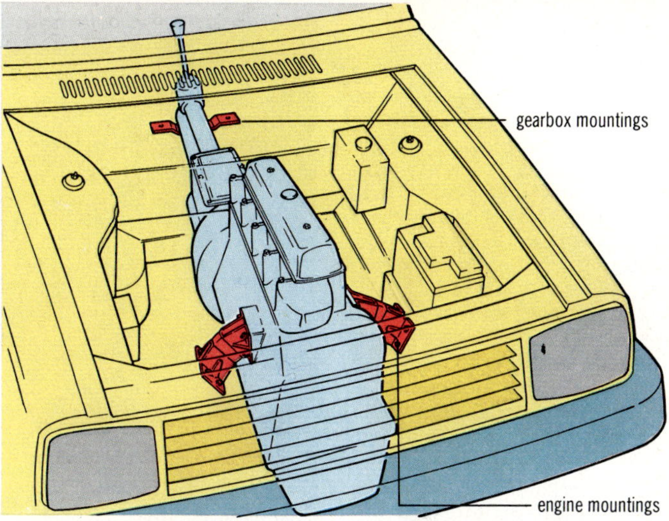

1. Typical locations for engine and gearbox mountings

Strange vibrations or juddering can be caused by a wide variety of faults. The most common cause of judder is a worn or badly adjusted clutch so suspect this first. You can usually spot a faulty clutch if the vibration is worst when you are starting from rest. Here you may even be able to feel the engine shaking about as you pull away. This sort of judder is common on front wheel drive cars, as there is not much give in the driveline.

Failure of an engine or gearbox mounting is another source of judder. In this case the judder is caused by the engine or gearbox actually moving as the drive is taken up. Have a look at the mountings (**fig 1**) to see if there are signs of oil contamination or wear. Old mountings can usually be spotted as the rubber peels around the edges. A quick if rather brutal test is to accelerate really hard in a low gear and then take your foot off the throttle sharply. If you hear a thud against the transmission tunnel you can be sure that the rear mounting needs to be replaced.

Most front wheel drive cars have an engine steady rod or stabilizer designed to stop the engine moving about while the clutch is being engaged. If the rubber bushes at each end of the rod get soft or break up, this will allow the engine to move about and so cause judder.

On the rear axle of many cars like the later Cortinas and the early Cavaliers, the rear axle is controlled by four links. These also have rubber bushes at each end which tend to break up after a time. When this happens, the axle moves sideways with a loud thump.

Although ignition and carburation are top of everyone's check list when a car won't start, they are not normally thought of as causes of vibration. Yet a loose LT or HT lead, which only comes free at high speed because of road and engine vibration, will cause the engine to misfire, run out of balance and shudder quite violently.

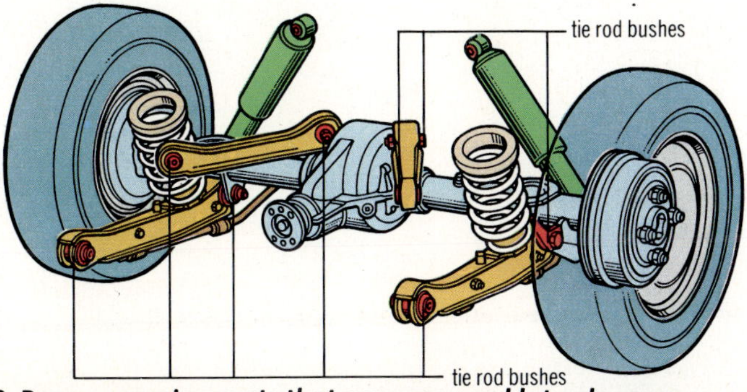

2. Rear suspension parts that may wear and let axle move

Similarly an out of tune carburettor may cause an engine to falter under acceleration giving a similar juddering sensation. Both these faults can often be identified by revving the car hard at a standstill. If the engine misfires and shakes, you have probably found the problem. Otherwise a quick check through both the ignition and fuel systems should reveal the cause of the fault. In particular, make a careful check of the carburettor.

One of the most common causes of transmission judder, particularly on older cars which have covered a high mileage, is a faulty universal joint (UJ). These can rattle, knock or produce 'wind up' in the drive depending on the type of joint and where in the transmission it is positioned. In most cases checking UJs is a very straightforward job. Replacing a faulty joint too, is generally quite easy, but a growing number of cars are being fitted with non-serviceable joints. On these cars a new propshaft or drive shaft will be needed. This too is a straightforward job which you can do yourself — but will cost considerably more.

STEP 2 CHECK THE UNIVERSAL JOINTS

To find and check the UJs on your car it is best to raise the car as high as you can on axle stands and ramps. But, because you may end up working at both ends of the car, use your stands, or your ramps as stands, and raise one side of the car. If you have a couple of old wheels handy, you can support one end of the car on ramps or stands and the other on the old wheels.

Chock any wheels still on the ground and slide your spare wheel on blocks of wood under the lowest part of the car to protect you should a stand give way. Put the car in gear and check that the handbrake is on. Now lie under the car and using an inspection lamp work your way along the transmission from the gearbox to the rear wheels until you have located all the drive train UJs.

When you have found out how many and what type of UJs your car has you can check them for wear. Again, work from the gearbox back as the first or second propshaft joints are the ones most prone to wear. Note that on some cars the propshaft UJs are hidden under a rubber or metal shroud designed to protect them from dust and dirt. These covers are usually secured in place by Jubilee-type clips.

To check a Hooke's joint first examine it visually. Any signs of rust probably indicate that the

1. Checking that the bolts holding the flanges are tight

FACT FILE

Location and types of propshaft UJs

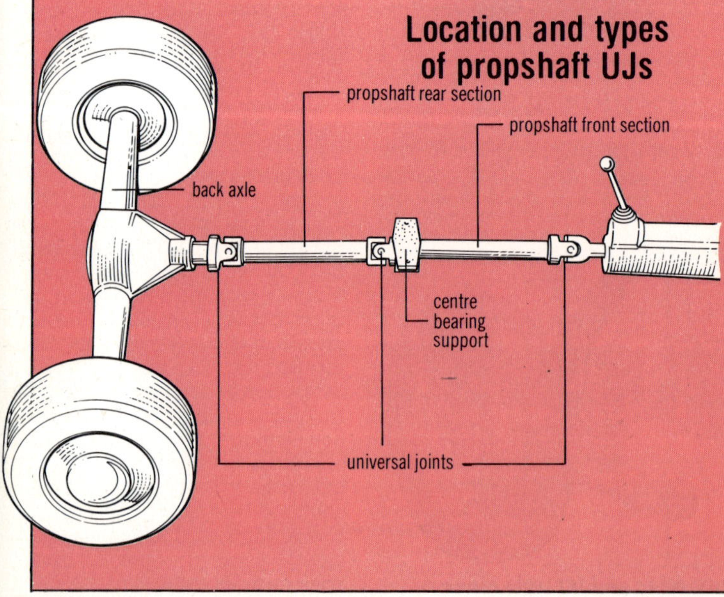

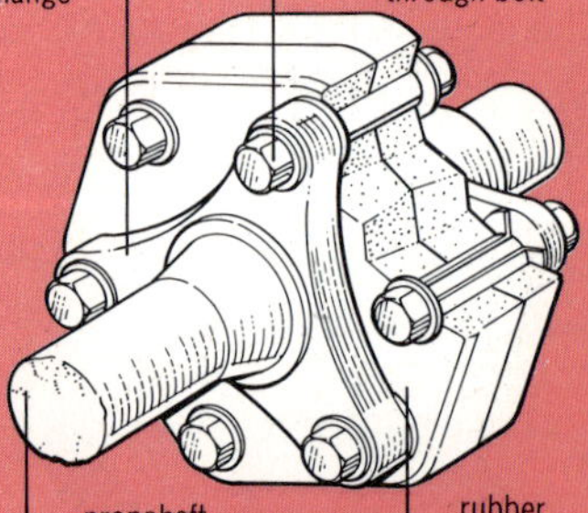

On front engined rear wheel drive cars universal joints are commonly found in three places: at both ends of the propshaft, at the centre of the propshaft and at each end of the driveshafts on cars with independent rear suspension (IRS). To complicate matters further, three different types of joint are widely used: the constant velocity joint, the

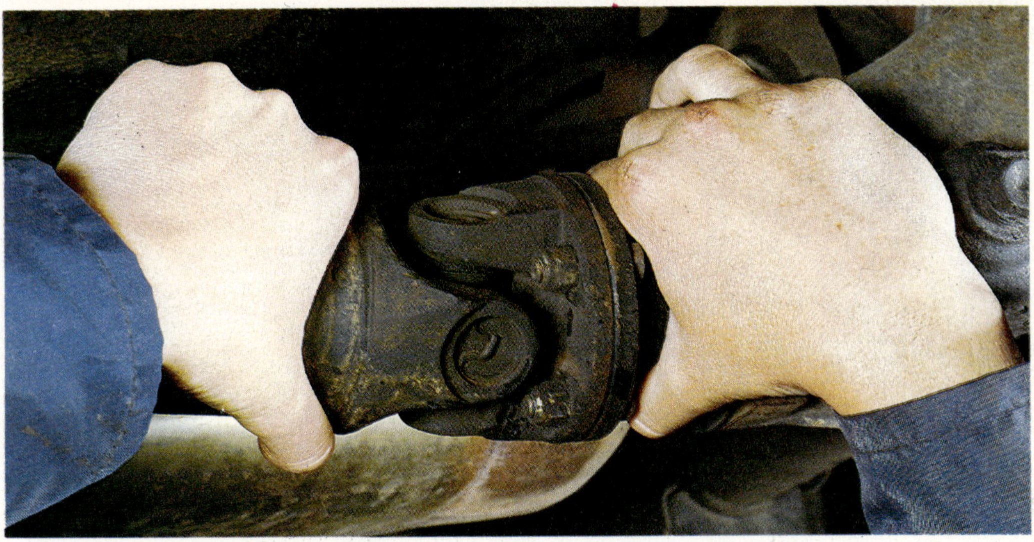

2. Checking for play in the joint. There should be no movement at all

joint needs renewing. Look closely at the bolts holding the joint (**fig 1**). Often these will be covered by a thick layer of dirt. Loose or flaking mud could indicate that the bolts are loose

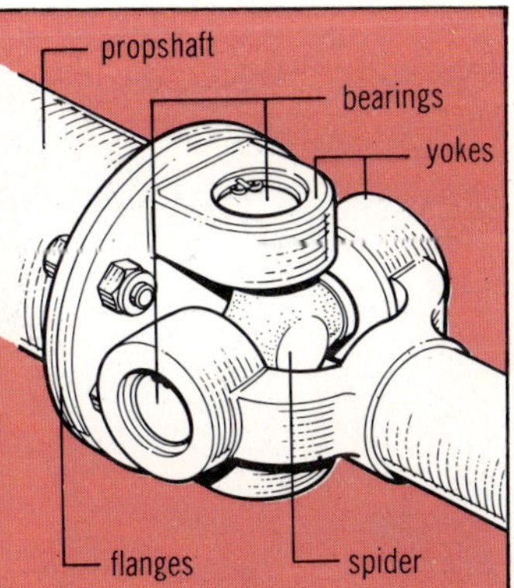

Hooke's joint and the rubber doughnut. The Hooke's joint, more commonly known as the Hardy Spicer joint, is by far the most common but the rubber doughnut is becoming increasingly widespread and is popular with many Italian manufacturers, particularly Fiat and Alfa Romeo who use them on the gearbox end of the propshaft.

— a common source of judder particularly when the propshaft has been disconnected and then reconnected using the same nuts. If the nuts are loose, tighten them up and see if the knocking noise has gone. Note that propshaft joints are held by special locking nuts, often with nylon inserts, and these nuts must be renewed if they are undone as they will only lock properly once.

To check for wear inside the joint, hold both flanges and twist them in opposite directions (**fig 2**). There should be absolutely no movement in the joint. If there is, you will need to renew the joint. When checking propshaft joints be sure that you are actually twisting the joint and not raising it — one end of the shaft has a spline where it fits into the gearbox and this will move if the shaft is raised. At the other end it is possible to be confused by backlash in the differential — so watch the joint closely as you twist.

When you are checking for wear in a driveshaft joint try applying the handbrake to lock the outer flange to the roadwheel and then lever the driveshaft with a large screwdriver and watch the joint for any sign of movement. You can also try spinning the wheel and listening to the joint and twisting the joint flanges in the same way that you would test a propshaft joint. If you hear any squeaks from the joints or feel any play, you can renew the joint in much the same way that you would do on a propshaft.

If your car has rubber doughnut UJs, first check them visually. Look for signs of oil contamination, loose bolts, perishing of the rubber or tearing around the metal inserts. This type of joint seldom causes severe judder, however, unless it is very badly worn. More often it produces a delay or vagueness in the transmission when the clutch is let in commonly called *wind up*.

If a joint in the propshaft needs to be renewed, you have to take off the propshaft. When renewing a rubber doughnut UJ you can sometimes leave the shaft in position but it is often easier to remove the propshaft. If a driveshaft joint needs renewing, it is usually easier to take off the shaft.

STEP 3 REMOVE THE PROPSHAFT

1. Marking the flanges so that the joint can be reassembled in the same position

Propshafts are carefully balanced by the manufacturer so that they do not vibrate when spinning at speed. To make sure that this balance is not lost you must replace the propshaft in exactly the same position as it was in before you removed it. To do this, first mark each section or flange of the shaft with chalk (**fig 1**) or a sharp scribing tool so that you can line up the marks again later.

Start work at the differential and undo the nuts holding the UJ flange to the rear axle (**fig 2**). Turn the shaft so that you can reach the nuts in turn. The weight of the propshaft may immediately break the joint — so do not lie with your face under it — if not, try pushing the shaft towards the gearbox or tapping the flange with a mallet and twisting it.

Now unbolt the flange at the gearbox end and lift the shaft away from the car. If the propshaft fits into the gearbox on a sliding spline lift the shaft carefully away from the gearbox and have a drip tray handy to catch any oil.

If your car has a two piece propshaft and you are only renewing one of the joints, it may be worth splitting the shaft. To do this, mark the position of all the flanges and then unbolt the gearbox flange. Next unbolt

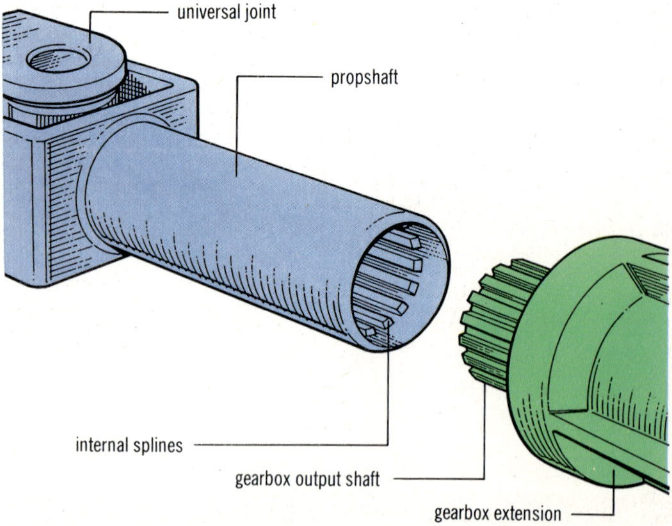

3. How gearbox and propshaft splines fit together

2. Removing the bolt from the flange of the joint

the rear part of the propshaft from the centre bearing. The bracket holding the centre bearing to the car, can now be undone. Lift away the shaft and centre bearing together.

To remove the rear part of the shaft, simply unbolt the flange from the differential.

In many cases, however, the job is easier if you simply remove the whole propshaft. If you do this and the two parts of the shaft are splined together in the middle try to keep the two shafts together. This will avoid any difficulties matching up the joint later.

If the universal joints are held together with circlips, you can easily fit new ones as described in the next step. But if your propshaft is fitted with the other type, there are a number of companies that will exchange your old shaft for another. This is much cheaper than buying a replacement.

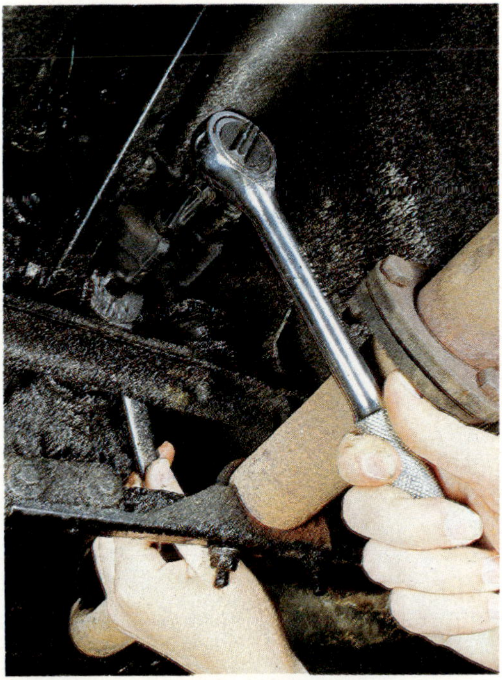

4. Undoing the doughnut joint through-bolt

5. Unfastening the centre bearing support

STEP 4 OVERHAUL A HOOKE'S JOINT

Universal joints which can be serviced are held in their yokes by circlips (**fig 1**). These will either be clearly visible in each eye of the yokes or hidden on the inside of the yoke. Internal circlips can usually be removed with circlip pliers. External circlips must be prised out of their grooves with a suitable tool such as an old screwdriver. Circlips can be both brittle and springy so it is a good idea to wear some form of eye protection when you remove them.

If the circlips are rusted in place, break them up and pull them out with pliers.

When you have removed all four circlips, support the UJ firmly across the top of an open vice. Now tap the flange, near the yoke, with your mallet. Keep tapping it until the UJ bearing cup rises up out of the yoke (**fig 2**). When it has come far enough for you to get hold of it, turn the joint over and tap the other side of the flange. Then pull out the bearing cups with a Mole wrench or by nipping them in the jaws of the vice and tapping the yokes from below. You will now have split the joint and the remaining bearing cups can be removed in exactly the same way. If tapping the flange does not make the bearing move, put the joint in a vice. Put a small socket, that will fit through the eye of the yoke, against one bearing. Place a large socket, through which the bearing can pass, against the opposite bearing. Now tighten the vice until the bearing moves out of the yoke.

When the joint is completely dismantled you should have four circlips, four bearing cups with oil seals and needle rollers, the spider and the two yokes. Clean the yokes with a degreaser such as Jizer then examine them closely for wear. If the eyes in the yokes are severely elongated (**fig 4**) then the yoke should be replaced. This may mean renewing the propshaft if the yoke is part of the shaft. As propshafts are expensive it may be worth trying to find an engineer who can replace the yoke by building up the yoke with weld and then reboring the eyes.

Now check your overhaul kit to make sure you have bought the correct parts. The joint will normally be supplied ready assembled. Some manufacturers, notably Datsun, supply up to 12 circlips or 'snap rings' of different thicknesses. These allow the joint to be packed out exactly and eliminate any play

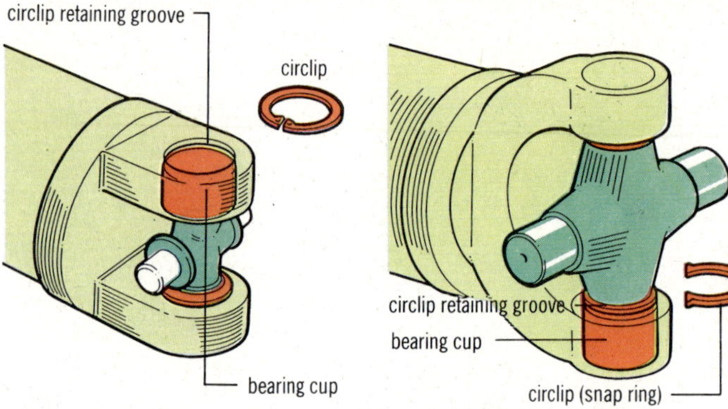

1. How circlips hold the bearing cups in place

2. Tapping bearing cup free

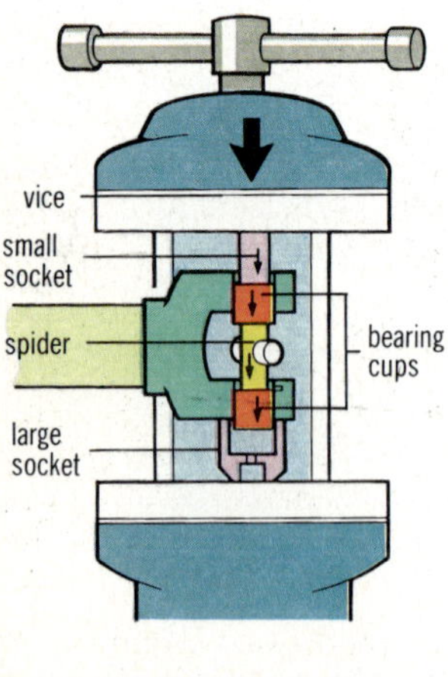

3. Driving bearing cup out

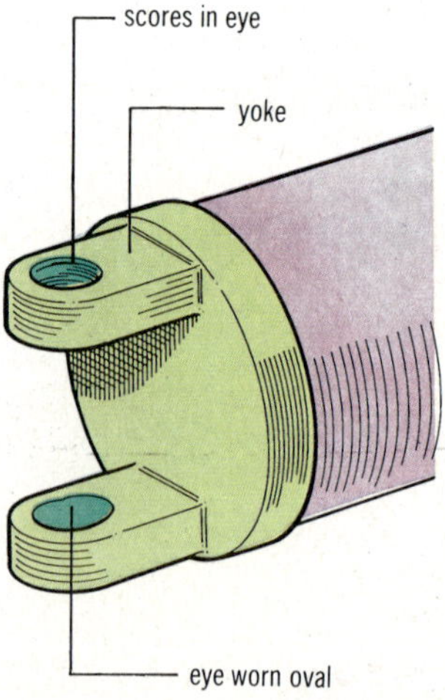

4. Signs of a worn yoke

5. Fitting new bearing cup in place on spider

6. Driving bearing cups home with vice

7. Fitting circlips to locate bearing cups

caused by poor tolerances.

To fit the new joint first remove all the bearing cups from the new spider, taking care not to lose any of the needle roller bearings. Fill each cup one third full with grease. Now insert the spider into the smaller of the two yokes and push it hard up against one eye of the yoke so that one of the bearing faces of the spider sticks out of the yoke. Slip a cup over the spider and use your vice to press it home (**fig 6**). Make sure that the bearing cup slides straight on the spider. When the cup is flush with the yoke, fit a small socket between the jaw of the vice and the cup and force the cup partly through the eye so that the bearing surface of the spider protrudes through the opposite eye. Then fit a cup on to that 'leg' of the spider and press it home in the vice.

If the joint has internal circlips, fit one before you push the second cup home. This will make sure that the first cup seats securely against the clip. You will then be able to fit the other clip into place. If the circlips are external, try to centralize the spider in the yoke and then fit the clips. If you have a set of clips of different thicknesses fit the one that seats most tightly against the cups. When you are pressing the cups into place you may meet a problem if the yokes are splayed. In this case you will have to pack the jaws of your vice with wedges of wood or metal to ensure that the cups are pressed in squarely.

Once you have fitted the bearing cups to one yoke, repeat the whole operation for the other one. When the joint is assembled check to see if it moves freely. If the joint seems a little tight, lay the joint over your vice as you did when you were trying to dismantle it and tap the yoke with a mallet. Then turn the joint over and tap the other yoke. This should ensure that each cup is seated correctly and that the joint moves freely.

You can now refit the propshaft or driveshaft to the car. Take care to line up the marks you made on the flanges and use new locking nuts.

OVERHAUL A DOUGHNUT JOINT

1. Unbolting doughnut joint from propshaft flange

2. Compressing old doughnut joint with hose clips

3. Fitting new doughnut joint — note metal band

The rubber doughnut universal joint, is becoming increasingly popular and is commonly found at the gearbox end of the propshaft.

Because it is simply held by through-bolts — usually six — you can change the joint without removing the propshaft. In some cases, however, the job is much easier if the shaft is removed first (see Step 3).

To get the joint off, simply unbolt the through-bolts which pass through the metal inserts in the joint to secure the doughnut to the driveshaft flanges (**fig 1**). But before you take the joint out mark the flanges so that they can be reconnected in the same position when the new joint is fitted.

When you buy a new rubber doughnut joint it will be supplied fitted with a metal band crimped in place. This band keeps the rubber compressed so that all the metal inserts will line up with the bolt holes in the flanges of the propshaft. *Do not remove this band until the joint is in place.*

Slip the joint between the flanges, noting any marks you made earlier, and then bolt the joint in place using new locking nuts (**fig 3**). When all the nuts are secure, cut the metal band with tin snips, taking care not to damage the rubber.

If for some reason you have removed a rubber doughnut joint and then wish to replace it you will find it very difficult to align the flange bolts with all the metal sleeved holes in the joint because the joint will have expanded. Try making a compression ring out of two of the largest Jubilee clips you can find (**fig 2**). Undo both clips completely then fasten them together to make one extra-large clip. Slide the clip around the joint then tighten the clip to force the joint into shape.

When handling and fitting rubber joints, keep them free of oil or grease.

Replacing track rod ends

Does your car's steering seem a little vague, or are the front tyres balding along one edge? A worn track rod end is the most likely culprit

The track rod ends are vital steering components, but all too often they go unchecked until the steering is dangerously worn and the car fails the annual roadworthiness test.

When to do this job
If the steering has play in it
If a tyre starts to wear along one edge of the tread

What this job involves
Breaking the taper joint
Removing adjustable track rod end
Removing fixed track rod end
Fitting a new joint

Points to watch
Sometimes the taper joint on a track rod end will come loose before you have undone the nut. If this happens, cut the nut off with a small hacksaw
Always take your car to a tyre specialist to have the tracking checked when you have fitted any new parts to the steering

To do this job
Tools: Ball joint separator; spanners; torque wrench; axle stands; paint brush (maybe); hammer (maybe); wheel brace; jack; bricks and length of wood (maybe: see Tip — Force fit)
Materials: New joint if necessary; paint (maybe); penetrating oil; new split pin (maybe)
Time: As long as all the nuts come free, and you use the proper separating tool, no more than two or three hours at the very most
Degree of difficulty: This job should not cause any real problems but care must be taken not to upset the wheel alignment (tracking)

If you have the job professionally done . . .
Is the steering precise and not binding? Is the new joint clean and the rubber boot intact? Has the wheel alignment been checked and adjusted?

STEP 1 IDENTIFY THE JOINT

The track rod end, often called a ball joint, transmits movement from the steering rack, or steering box, to the front hub. It is basically a ball and socket joint (**fig 1**) designed to allow the wheel hub to move up and down and from side to side, independently of the rest of the steering assembly. There are two types of track rod end: adjustable and fixed.

The adjustable joint (**fig 2**) is the more common, and is a separate casting which screws to the end of the track rod or steering rod. On a rack and pinion system there are two joints: one on the end of each track rod (**fig 3**). Where a steering box is fitted there are four or six joints, depending on the type of system fitted to your car. **Fig 4** shows the most common positions of the joints,

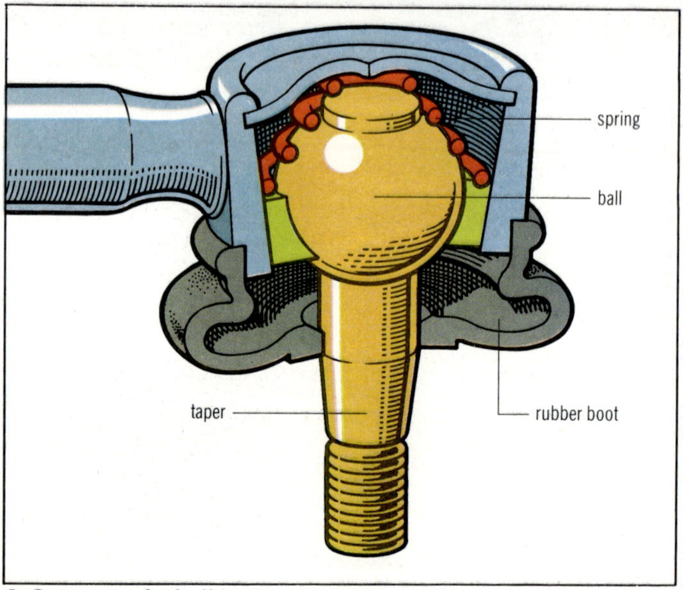

1. Cutaway of a ball joint showing major components

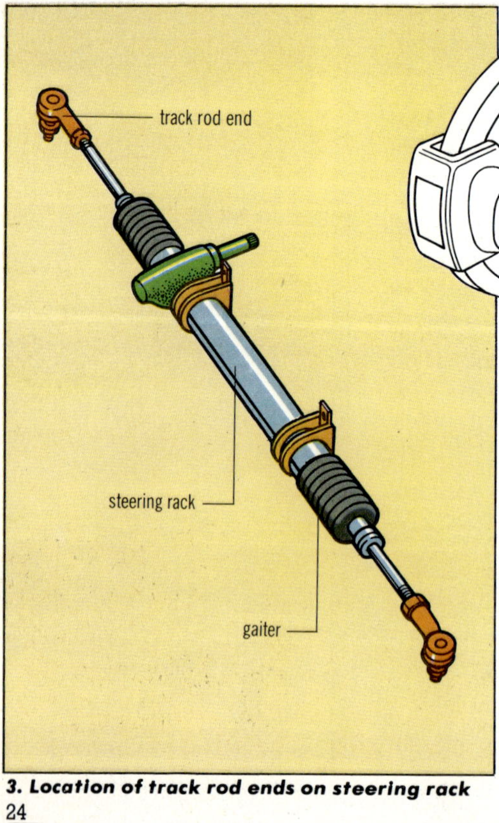

3. Location of track rod ends on steering rack

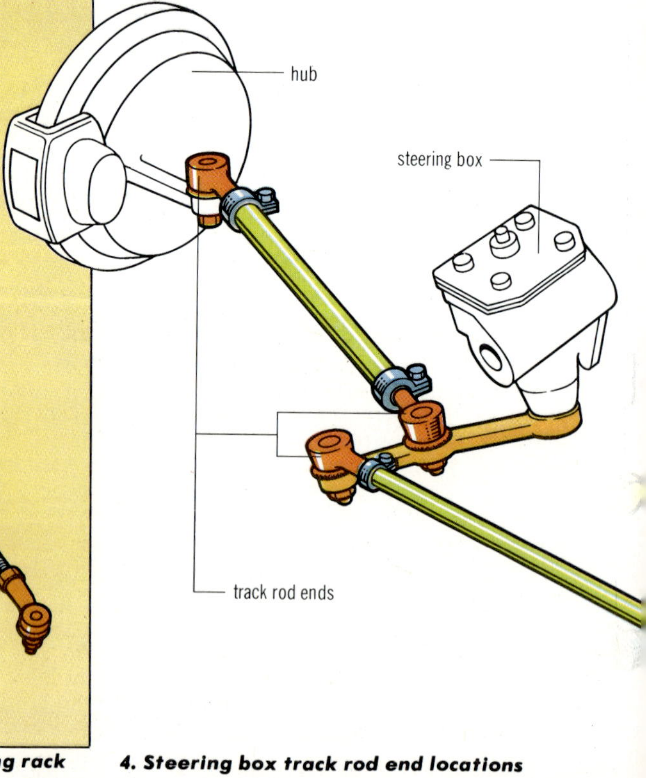

4. Steering box track rod end locations

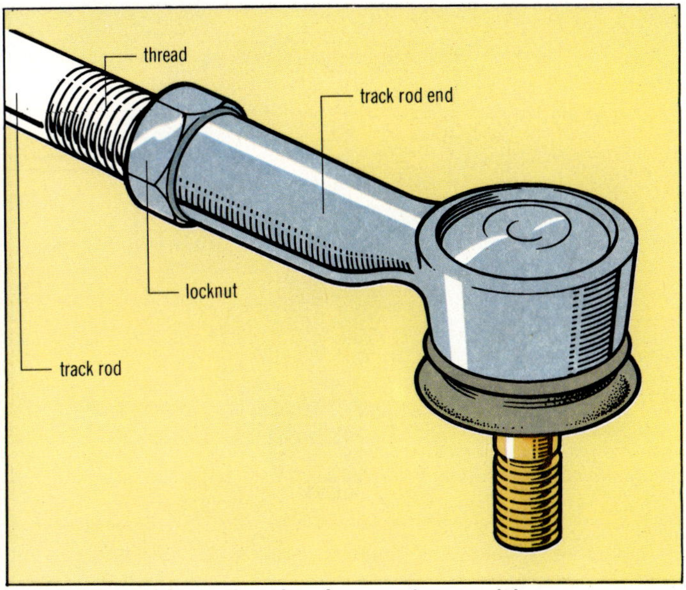

2. The adjustable track rod end screws into position

but it is possible that the steering system on your car may look slightly different. The joint is kept locked in position by either a locknut or by clamps depending on the design. See Step 3 for how to remove the joint, once the taper joint is released (Step 2).

The fixed track rod end (**fig 5**) is less common than the adjustable type and, in most cases, is usually fitted to only one side of the car. The other side will have a normal adjustable joint.

The fixed joint is, in effect, both ball joint and track rod in one unit. There is a normal ball joint at the wheel end, while the other has a rubber bush set into it. A bolt goes through the centre of this bush and fixes the whole rod to the steering rack. On a few cars the bolt is secured by a flat metal tab washer which

5. Fixed track rod end with alternative type

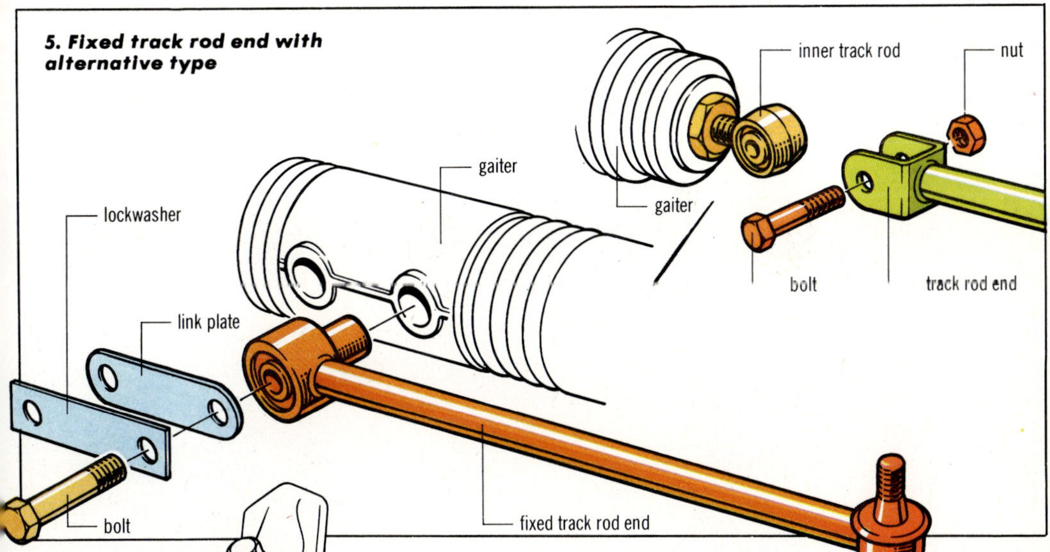

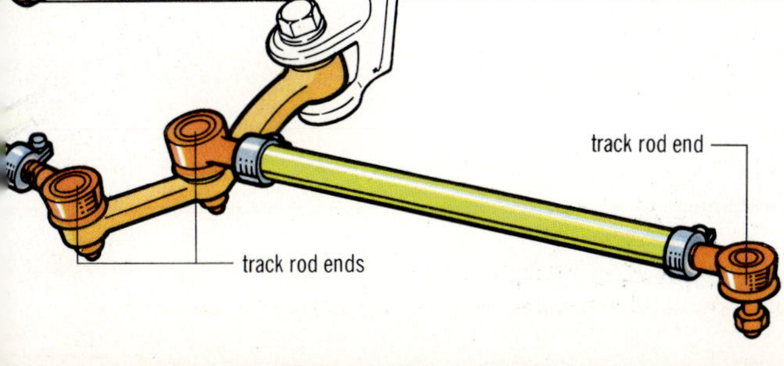

needs renewing every time you undo the bolt. Step 4 deals with replacing this type of joint, after the taper joint is undone. Track rod ends are now greased for life so try to keep the rubber boot in good condition.

STEP 2 RELEASE THE TAPER JOINT

First, raise the car up as high as possible and support it on axle stands. You must make sure it is securely supported as you may rock the car later on as you remove the track rod ends.

All track rod ends are fixed at the wheel hub end by a tapered joint (**fig 1**). This is designed so that when the nut on the top of the joint is tightened, it draws the tapered part up into the tapered hole in the steering arm.

This locks the taper into the arm so that even if the nut comes loose the taper will still hold the joint together.

The taper will be a very tight fit in the steering arm, especially if the joint has not been disturbed for some time.

To release the joint, start by loosening the nut on the top. This is either a self-locking nut or a castellated nut which is held in place by a split pin. If your car has the castellated sort, give it a good squirt with penetrating oil, then use pliers to straighten the ends of the split pin and pull it out — if it refuses to budge, cut the head off with a hacksaw, then use a hammer and a small drift to knock it out. If you do not have a drift, file the end of a thin nail flat and use that. Remember to buy a new split pin unless you are renewing the joint — new joints

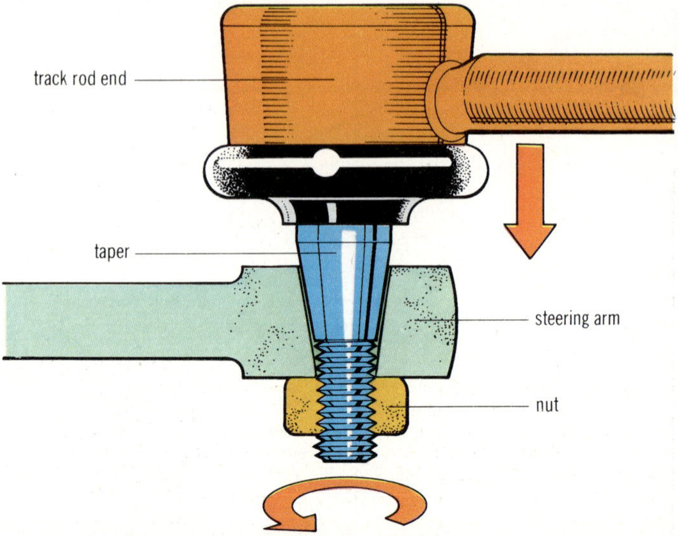

1. As nut is tightened, taper is drawn into steering arm

2. Splitting the taper joint using a double arm separator and large spanner

usually come complete with a new self-locking nut. Loosen the nut on the joint but do not take it right off just yet.

You now have to break the taper joint. Although it is possible to shock the joint apart using two hammers on either side of the joint, you run the risk of breaking the steering arm or part of the steering rack or box, so it is best to use the proper tool (see Fact File — Ball joint separators).

Once the joint is free, go on to the step which deals with releasing your type of track rod end from the steering.

FACT FILE

Ball joint separators

You have a choice of several ball joint separators which can be bought or hired quite cheaply. The cheapest are the double wedge sort which are hammered between the steering arm and the ball joint from opposite sides. As the two wedges are hammered inwards they move over each other, effectively expanding, forcing the joint apart. The shocks from the hammer blows also help shock the joint free. You can also get a single wedge separator but these are bulkier than the double wedge sort and require more force to split the joint effectively.

The disadvantage with wedge splitters is that they tend to destroy the rubber dust boot. If the joint is to be changed this is no problem as the joint comes complete with a new dust cover, but if you are undoing the joint to remove the hub for example, then it is better to use a screw-type separator. These come in two sorts, the box sort and the double arm sort. These both have a forked part which is forced between the track rod end and the steering arm. The bolted part is then screwed up to bear on the track rod end nut and so split the joint. If you do accidentally split the dust cover and cannot get hold of a new boot from your dealer you can try repairing the old one with a plastic rubber adhesive available from hardware shops. Bear in mind that this is only a temporary measure and the joint will need replacing later on as it will wear quickly.

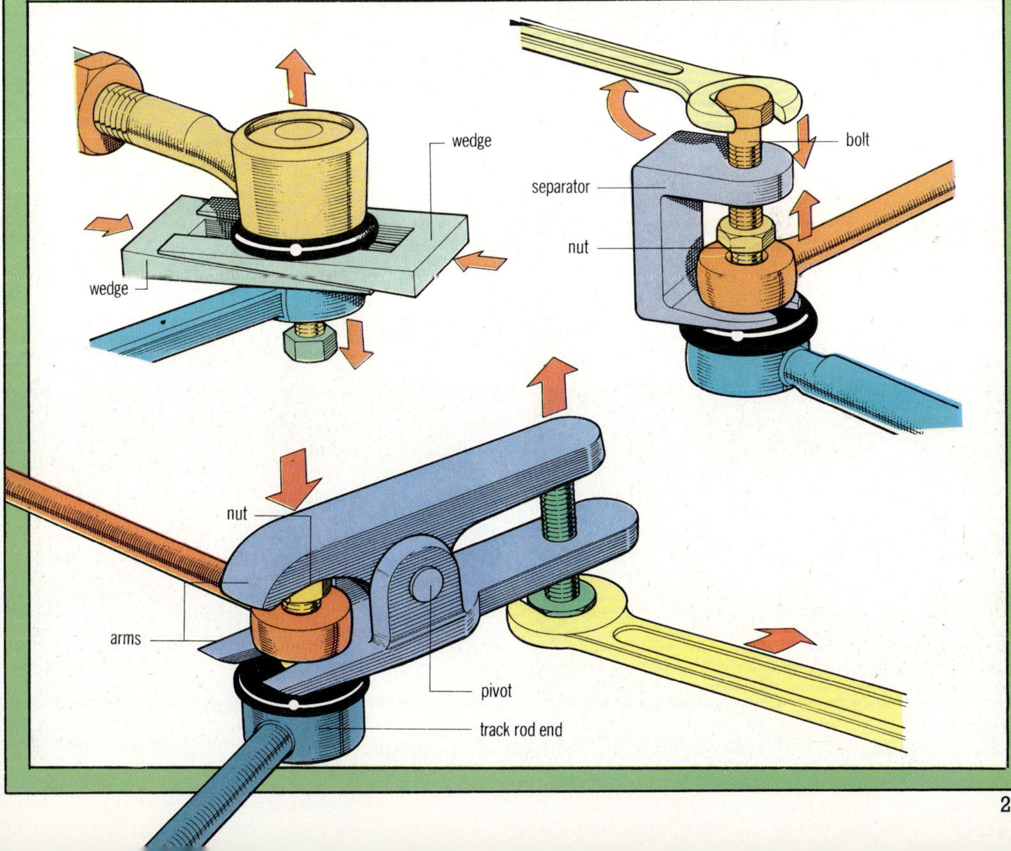

STEP 3 REMOVE ADJUSTABLE TRACK ROD END

To release the ball joint from the track rod, you must first release the locking device. This can be a locknut which bears on the end of the joint, or two clamps which squeeze a threaded pipe on to the joint (**fig 1**).

First give the locknut or clamp a good dose of penetrating oil. On a joint with a locknut, use two spanners to undo the locknut; one on the nut and one on the joint itself. If your joint has no spanner flats cast on it, use Mole grips on the joint. The clamped joint simply needs the clamping bolts undone — it is no good undoing just one, both must be slackened to remove the joint.

On some cars fitted with a steering box, instead of a steering rack, the joints on the left-hand side of the car may have a left-hand thread. These undo in the opposite direction to normal, so if the locknut shows no signs of moving turn it the other way.

Carefully loosen the locknut or clamp bolts, but do not move the joint just yet. Now mark the position of the joint in relation to the track rod, using a blob of paint. This will help you get the new joint in the same place as the old one and will avoid

1. Two joint locking devices

STEP 4 REMOVE FIXED TRACK ROD END

Once the taper joint has been released, you can go on to free the track rod end where it is fixed to the steering rack.

First knock back the metal tab washer (if fitted) using a hammer and a large flat-bladed screwdriver or chisel (**fig 1**). Using a snug fitting socket or spanner, undo the bolt holding the track rod end to the steering rack (**fig 2**).

Where the inner bush fits into the steering rack gaiter be ready to cover the hole when the rod comes away as oil will spill out of the steering rack (see Tip — Bung it up). With the bolt out, pull the whole assembly out of the rack being careful not to damage the gaiter.

You may find that it is possible to replace the one-piece rod with a two-piece type (**fig 4**). You will need to buy a new inner and outer rod (like the ones fitted to the other side of the car) and find the adjustable threaded part and the locknuts at a scrapyard. Some manufacturers even supply the complete adjustable rod as a service replacement. Put the whole lot together, then measure the old fixed rod and adjust the new rod to the correct length. You can now fit the rod to the steering rack (Step 5).

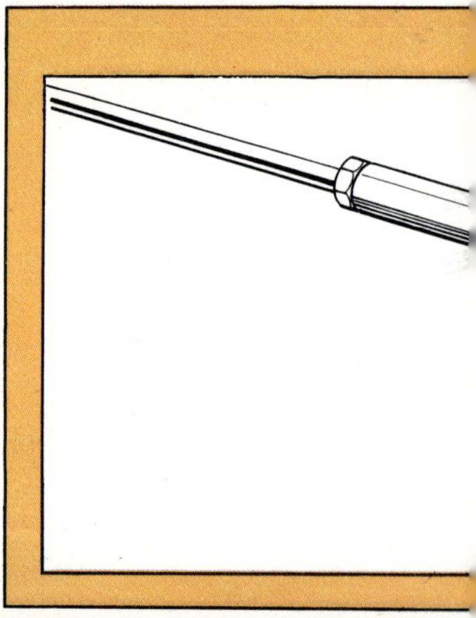

1. Knocking back tabwasher

2. Undoing the securing bolt

disturbing the steering alignment (tracking). Alternatively, use a ruler to measure the amount of thread protruding from the joint.

Once the locknut or clamps are slackened, use Mole grips to grasp the track rod while you unscrew the joint. As you remove it, count how many turns it takes — this will help you align the new joint later. Where the locknut has a left-handed thread, the joint has to be turned clockwise. If you have this type of joint, make sure that you buy a replacement with a left-handed thread.

2. Releasing the locknut

3. Marking track rod with paint

TIP

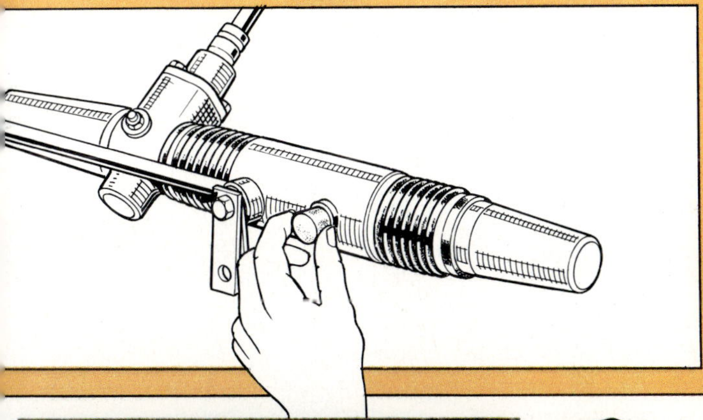

Bung it up

If your car has fixed track rod ends which bolt through the steering rack gaiter, oil will spill out of the rack when you remove the track rod end. To avoid losing too much oil, push a rubber bung (available from home brewing shops) into the hole in the rack gaiter. When the time comes to fit the new rod, pull the bung out and get the rod into position as quickly as possible.

3. Pulling rod out of rack

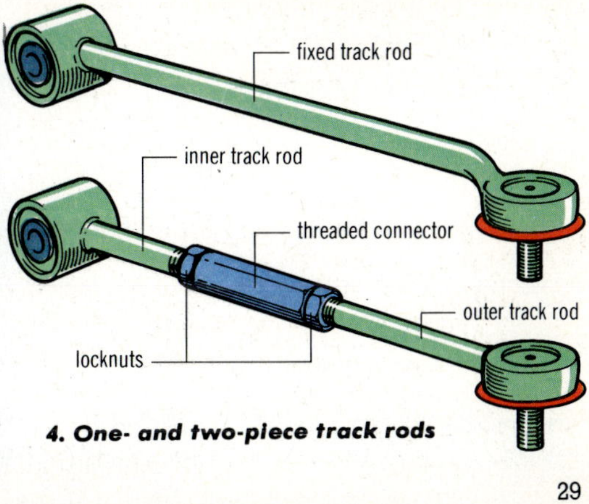

4. One- and two-piece track rods

STEP 5 REFIT THE JOINT

When you come to refit the joint it is simply a case of reversing the removal procedure. Put the joint back on, making sure the end lines up with any paint marks you made on the track rod (**fig 1**). On a fixed track rod end fit the rod into the rack and fit a new tab washer. Then, put the bolt in and secure it hand tight. Do not tighten up any locking nuts just yet — fit the taper joint to the hub first.

Take the new nut off the taper joint and swivel the hub so that the taper joint can be fitted into its hole in the steering arm. Fit the arm back on to the joint and tighten the nut up. It has to be tightened to a specific torque, which varies from car to car and should be quoted in your manual. If your own manual does not give this figure, your dealer will be able to.

If you do not have a torque wrench, this does not really matter. Use a ring spanner of normal length instead. Then do the bolt up as tight as you comfortably can with the ring spanner using one hand only. This will tighten the nut to roughly the right torque.

If you find that the joint goes round, as you try to tighten up the nut (this often happens with the self-locking type), you need to wedge the joint more tightly into the steering arm (see Tip — Force fit).

Once the joint is done up, tighten up the inner locknut, clamps or bolts, then bend over any tabwashers (**fig 2**).

Even though the joint will now be in its original position, the steering alignment may have been disturbed. To be on the safe side, have it checked as soon as possible, as incorrect alignment will quickly scrub the tread off one edge of the tyres.

Finally, refit the roadwheel and lower the car. Remember to tighten the wheel nuts fully once the car is on the ground. You can then road test the car.

2. Bending new tabwasher using Mole grips

TIP

Force fit

When refitting the taper into the steering arm, you may find that you cannot tighten up the self-locking nut without the taper turning round.

To force the taper further into its hole, you can use a jack (a hydraulic or scissor jack is best) or bricks and a length of wood.

If using the jack, position it directly under the joint then apply pressure to lock the taper up. If you do not have a suitable jack, use the wood as a lever, pivoting it on the bricks until you can hold the joint into the steering arm.

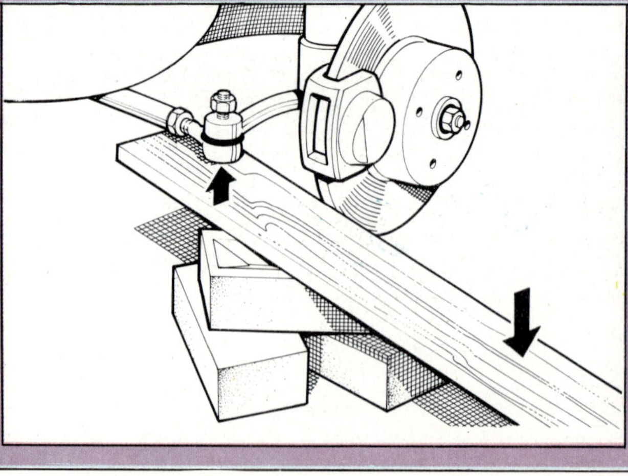

Front wheel bearing service

When the front wheel bearings rumble it means that they need attention. This may just be lubrication but new bearings may be required. Both jobs are straightforward

Faulty front wheel bearings can affect the safety and handling of your car as well as causing the front tyres to wear unevenly. In extreme cases, a wheel bearing fault can cause the entire wheel to lock up — or even fall off — while the car is in motion. For these reasons, the wheel bearings should be carefully checked as part of your car's routine servicing.

When to do this job
As recommended by the manufacturer or if you hear rumbling noises from front wheels and can feel excess play in the bearings

What this job involves
Checking wheel bearings
Adjusting wheel bearings
Removing wheel bearings
Lubricating wheel bearings

Points to watch
Kits are available for many cars containing a wheel bearing plus the oil seals required
Check the adjustment of new bearings after 200 miles

To do this job
Tools: Jack; axle stands; wheelbrace; screwdriver; pliers; hammer; hub puller (maybe); bearing puller (maybe)
Materials: New bearings (maybe); HMP (high melting point) grease
Time: Up to two hours per wheel
Degree of difficulty: Hub may be difficult to remove; bearings may be so tightly embedded that you will need to hire a bearing puller

If you have the job professionally done . . .
Do wheels turn smoothly and quietly? Has free play disappeared when you try to rock the wheels?

STEP 1 CHECK THE WHEEL BEARINGS

You can tell when the front wheel bearings are loose or worn by checking the hub, with the wheel attached to it, for free play. Some people try to do this with the road wheel still resting on the ground, but it is far better to make the check after the front of the car has been jacked up and supported with axle stands. With the wheel off the ground, grasp the top and bottom of the wheel and try to rock it back and forth (**fig 1**). If you can feel the wheel rock, check that this play is in the hub assembly and not looseness in the suspension or steering. You can do this by replacing your hands on the wheel in the three o'clock – nine o'clock position and

1. Checking for play in the wheel bearings

HOW IT WORKS

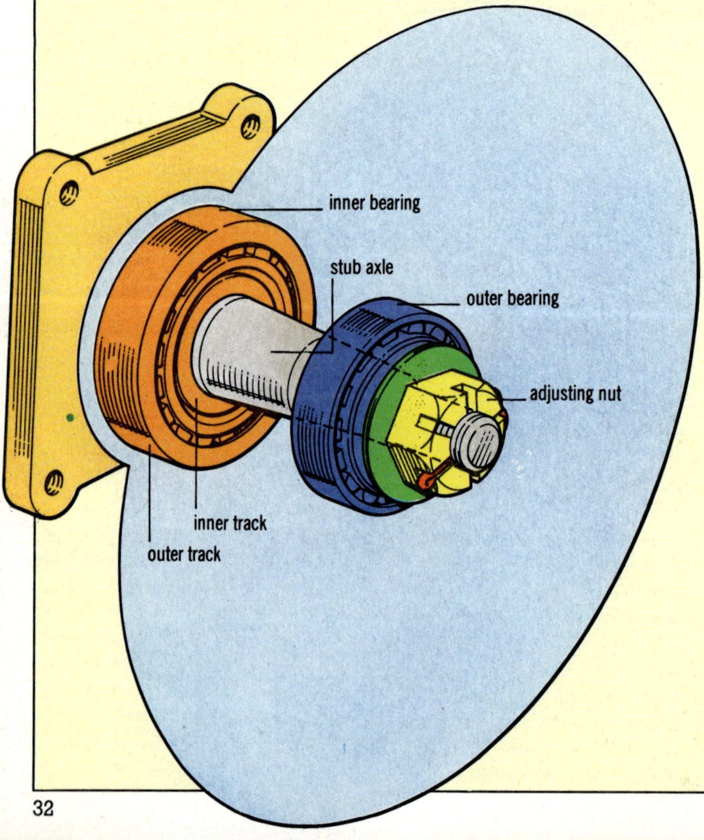

Wheel bearings

On most cars — whether front or rear wheel drive — the hub to which the front wheel is bolted spins on a stub axle. The wheel can turn because the hub is mounted on inner and outer bearings.

Depending on the type and make of car, these bearings may either be caged ball bearings or, more commonly, roller bearings which move against inner and outer races fitted to each side of the hub. Usually these roller bearings are tapered so that adjustment is possible. The bearings are packed with grease to keep them constantly lubricated.

The correct adjustment of tapered bearings is very important and, in most cases, this is a job which can be done by a DIY mechanic. Bearings that have either too little or too much clearance will rapidly wear out and will often start to make a rumbling or clicking noise which may be particularly noticeable when the car is cornering. Bearing noise is usually a sign that the bearing is badly worn and in need of renewal.

rocking it again. If there seems to be the same amount of play as before, then the wear is probably in the wheel bearings, though there may also be wear in the track rod end or the ball joints.

A small amount of play is allowed in most wheel bearings — about 1/16th inch at the most. Adjust the bearings first where possible, and then go on to find the rest of the wear.

While the front wheels are off the ground, spin them by hand and listen for any noises which could be caused by a faulty bearing. Very slight noise may be cured by packing the bearings with fresh grease. If the wheel does not spin freely, this may be because the brakes are holding it. On drum brake models the wheel can be freed by slackening off the adjuster on the brake backplate. With disc brake models, slight contact between the pads and the disc is normal, so you may have to lever the pads away from the disc or even take them out to check the bearings. Check the brakes afterwards if you have to do this ever.

If you find that the bearings are making excessive noise or seem to have far too much play, it is unlikely that adjustment or cleaning and greasing are going to offer anything more than a temporary cure. In these cases it is better to fit new bearings (see Step 3).

Some cars, like the Fiat 127 and 128 have non-adjustable bearings fitted right into the hub. You cannot service these yourself but you can still save money by unbolting the hub from the suspension yourself and taking the unit along to a main dealer to have new bearings fitted into place.

Many other cars have non-adjustable wheel bearings and you should not try to deal with the wear here by tightening up the hub nut. If you try this, it removes the running clearance and often causes the bearing to overheat and break down. If the bearings are worn you have to renew them.

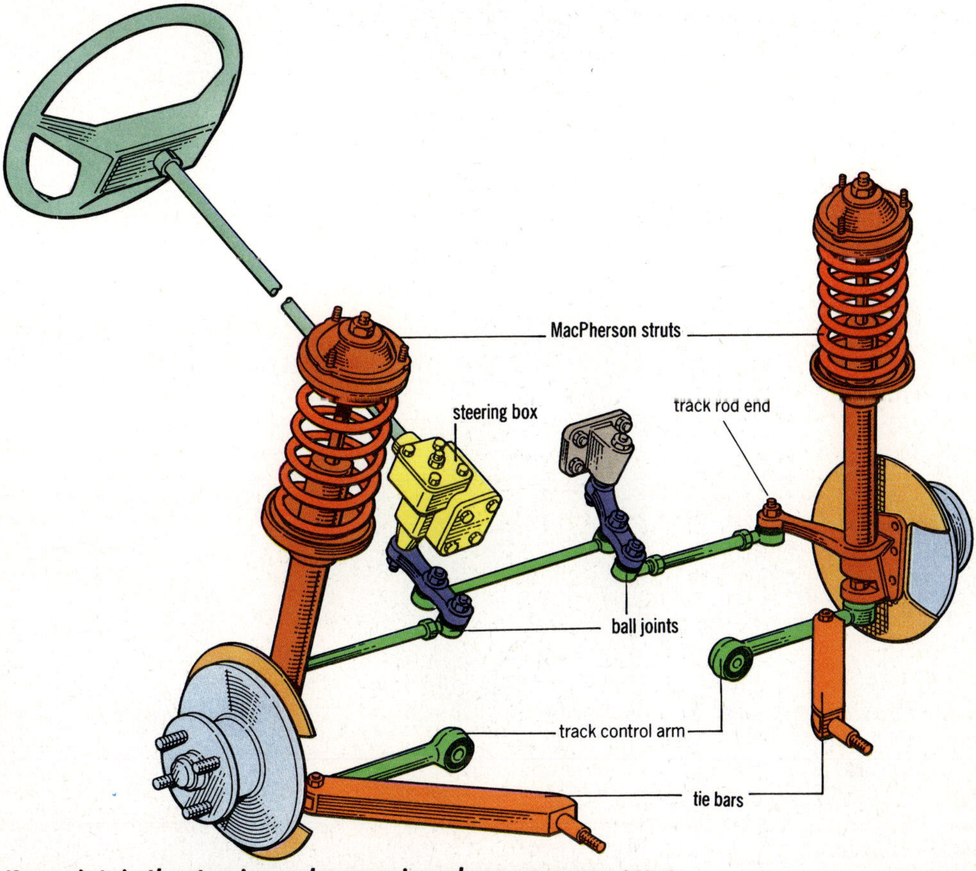

2. Key points in the steering and suspension where wear may occur

STEP 2 ADJUST THE BEARINGS

Small amounts of play in wheel bearings can sometimes be taken up by adjustment. The bearings and the hub assembly are usually held in place on the stub axle by a large nut or pair of nuts (**fig 1**). On some cars a single castellated nut is used together with a split pin which locks it in place. Other models use a second nut to lock the whole assembly in place.

Begin by consulting your manual to find out if your car is fitted with adjustable wheel bearings. Some are fitted with non-adjustable ball bearing assemblies which should give the right clearance as long as the hub nut is tightened to the correct torque figure.

If you find any play in this type of bearing you will have to fit a new one.

Once you are sure that you have bearings which can be adjusted, begin by loosening the nuts or bolts holding the road wheel in place. Slacken them only a little so that they can be undone easily once the car has been raised from the ground. Now jack up the front of the car and support it with axle stands so that the nuts or bolts can be removed completely and the wheel lifted away.

In the middle of the brake drum or disc you will see a domed metal cap. This holds the grease around the stub axle and prevents dirt and grit from getting inside the bearing assembly. Usually it can be removed by carefully levering it away with a large screwdriver or a pair of slip-joint pliers (**fig 2**). On some models the cap is threaded in place and can only be removed by turning it with slip-joint pliers — usually in a direction opposite to the rotation of the wheel — that is, clockwise on the left hand wheel. A few cars — notably the Volkswagen Beetle — have the speedometer drive attached to the grease cap on one wheel and the clip which holds it in place has to be

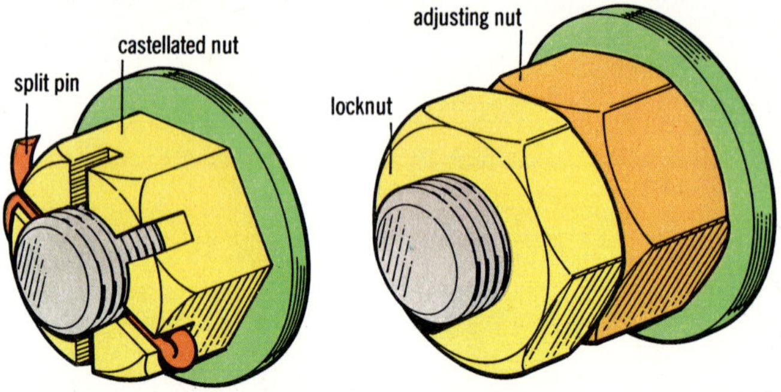

1. Alternative ways of locking the adjuster nut

2. Levering away the grease cap with a screwdriver

3. Loosening the central retaining nut

removed before the grease cap can be levered away. Store the clip carefully. When you have taken off the cap, you should be able to see the large retaining nut together with its locking device. If this consists of a split pin and a castellated nut, first straighten out the split pin with a pair of pliers and then gently withdraw the pin from its hole. If the split pin has broken off inside the hole and you cannot grasp it with pliers, use a hammer and a fine drift to tap the broken pin out.

On some cars the hub is held on the axle by two large nuts with a locking plate between them. Bearing adjustment is made by turning the inner nut, but before you can move it the locking tab which is hammered over the outer nut has to be straightened out using a hammer and a small cold chisel. Only then can you remove the outer nut and locking plate.

Another design, found on many Ford models, also uses two nuts which lock against one another. However a split pin is used instead of a locking plate to secure the nuts. Once you have pulled out the split pin you will need two similar spanners so that the inner nut can be held stationary while the outer nut is loosened. This done, the inner nut can be turned to make the adjustment.

Whatever system you come across, once you have freed the adjustment nut, it can be turned inwards to take up the bearing free play. But before you make the adjustment, take off the thrust washer and the outer bearing so that you can inspect them — they should just slide off the end of the stub axle. Check that the bearing looks intact and is well coated with grease — this should not be crumbly or waxy. If you find that the grease is not is good condition, you will have to dismantle the hub and bearing assembly so that you can clean the whole assembly and pack it with fresh grease (see Step 3).

If the bearing appears to be well lubricated, slide it back in place together with the washer and carry on with the adjustment. This procedure varies from car to car. Many DIY mechanics simply turn the large nut until most of the play has been taken up, but the hub still spins freely. This approach is rather crude, but it can give fairly satisfactory results with some cars. However, this should be regarded only as a temporary solution — it is far more sensible to follow the manufacturer's recommendations. Check your manual for the procedure. You may have to tighten the central nut to a certain torque figure and to turn the hub at the same time so that the bearing beds in evenly, and then slacken the nut by part of a turn.

However, a more common instruction is for the manufacturer to give an *end-float* figure. The end-float is usually measured by sliding a feeler gauge between the outer bearing and the washer. Slide a feeler gauge of the recommended thickness between the bearing and the washer (**fig 5**) — it should be a smooth sliding fit. If the feeler is a loose fit, or is very tight, use a spanner to turn the adjuster nut until the gap is about right. You may then have to turn the nut slightly one way or the other to enable you to refit the split pin.

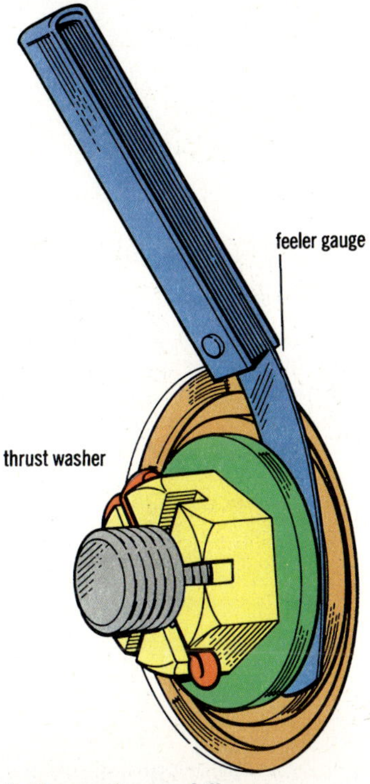

5. Measuring end-float

4. Removing the outer bearing from the hub

STEP 3 REMOVE THE WHEEL BEARINGS

If the bearings have worn too much for adjustment to cure the problem then they will have to be replaced. Begin by following the procedure in Step 2 for removing the grease cap and the large central nut. On disc brake models, before you can slide the hub and disc assembly off the stub axle you will have to unbolt the disc brake caliper and hang it out of the way. Often it is easiest to tie it to a nearby steering or suspension component with some strong wire. This will take the strain off the flexible brake pipe which can be left undisturbed. It is also a good idea to wedge a piece of wood between the pads to ensure that the brake pistons do not come out of position. Clamp and disconnect the brake hose if a rigid pipe stops you from lifting the caliper away.

As you dismantle the wheel bearing lay out the parts in the order you remove them on a clean sheet of paper. You may also find it helpful to make drawings as you go along to help you remember how everything fits together (**fig 3**).

Slide the outer bearing and thrust washer off the axle and then slide the entire brake disc and hub assembly away. In some cases a degree of force is needed to do this because the seal inside the hub will hold the assembly on to the axle. If you

1. Two bolts hold caliper

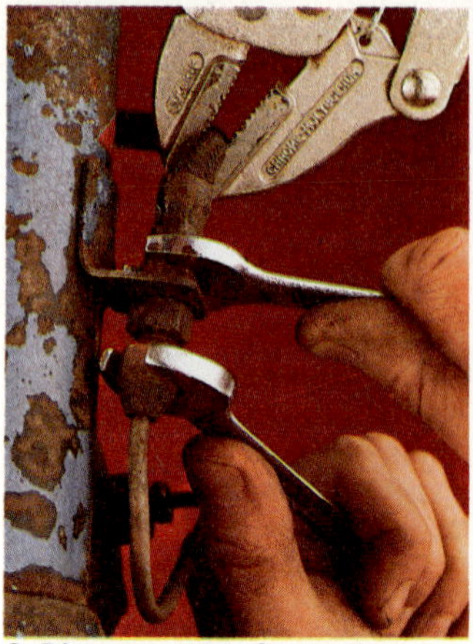
2. Disconnecting brake hose

FACT FILE

Hub pullers

A hub puller will be needed if the hub is so tight on the stub axle that you cannot pull it off by hand. The puller is first fastened on to the hub. Then the central bolt is tightened against the axle. As this bolt is turned, it cannot push the axle inwards because the axle is fixed. Therefore, it pulls the hub off the axle.

Most pullers have three or four legs but work on the same principle. Pullers of different sizes can be used for different jobs, such as removing tight wheel bearings.

Depending on the design of the hub, the puller may be fastened to the wheel studs or may reach behind the flange of the hub and pull on the flange.

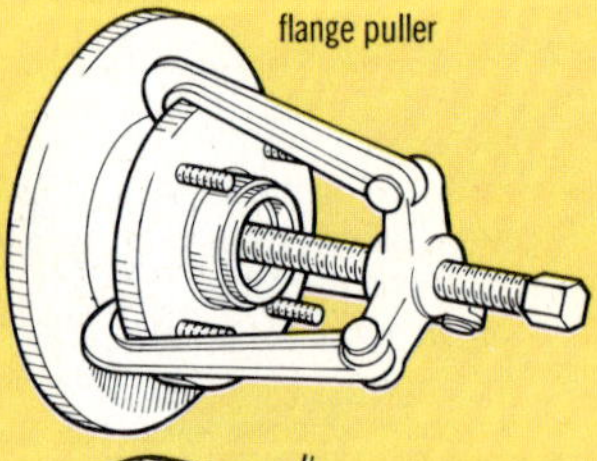

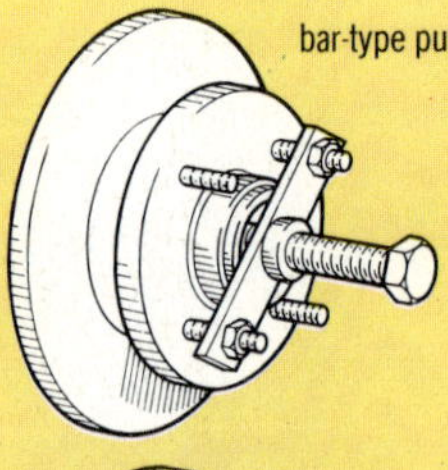

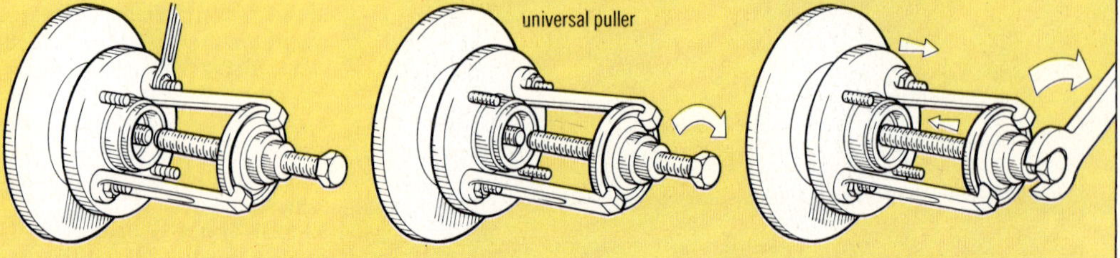

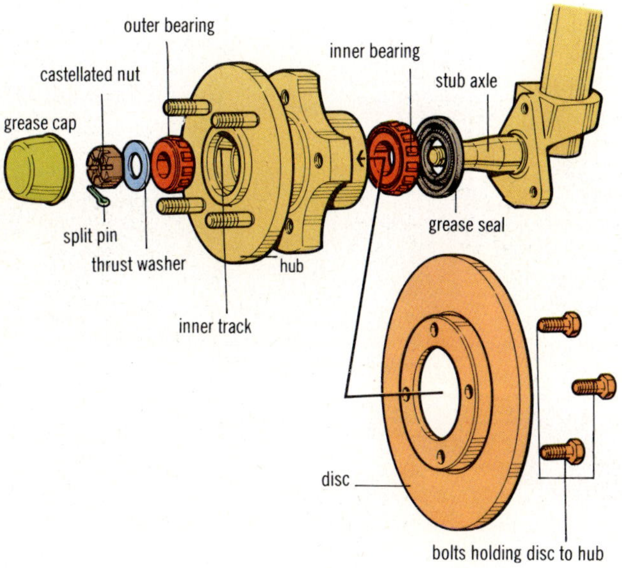

3. How the components on the stub axle fit together

fasten the road wheel back on with the nuts or bolts only finger tight you will probably get enough leverage to remove the hub from the axle.

If you have drum brakes you may need to slacken the adjusters first, but the procedure should otherwise be the same and the drum and hub assembly should slide away fairly easily. If you have a problem getting the hub off the stub axle — whether with disc brake or drum brake cars — you may have to hire an adjustable puller. Position the legs of the puller to grasp the back of the hub while the central column rests squarely on the tip of the stub axle. As the puller is tightened up the legs will draw the hub off.

Once you have got the hub off, use a screwdriver to prise the grease seal out of the back of the hub. This seal should always be replaced by a new one when you are ready to reassemble the bearings. If you are just going to pack the bearings with grease do not worry about removing the tracks from the stub axle. Just clean the old bearings with paraffin or petrol and dry them so that you can check them for worn rollers or general looseness. Check for signs of pitting or other damage on the rollers.

Now oil the bearings and see if they spin freely without any roughness. If you are satisfied that the bearings are still serviceable, work a quantity of high melting point (HMP) grease in between all the rollers making sure that the grease penetrates the whole bearing (**fig 4**). Pack the hub cavity half full of grease as well but no more or it will heat up and run everywhere.

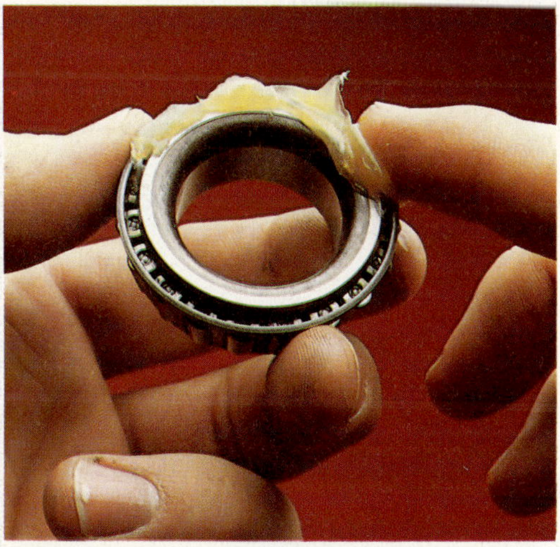

4. Packing new bearing with grease

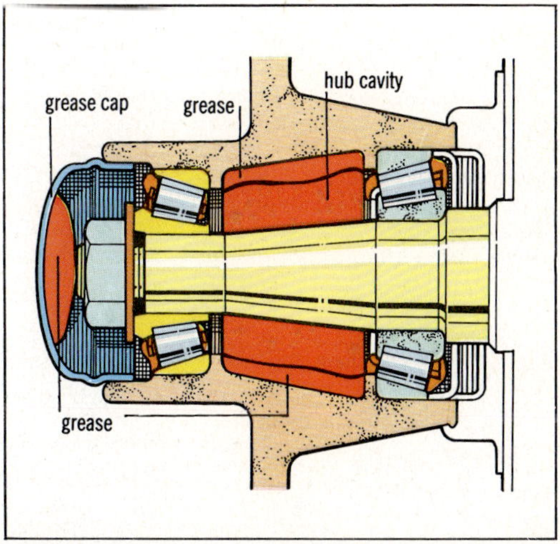

5. Where to apply new grease

STEP 4 FIT NEW BEARINGS

1. Removing inner track from hub

2. Using old track as drift

If you are fitting new bearings you will have to remove the tracks from the hub. These have to be tapped out using a hammer and a drift. To remove the inner one, carefully tap it out from the opposite side of the hub using the two slots which are usually cut into the hub for this purpose. Work from side to side until the track falls out and then repeat the job on the other track. You may need considerable effort and a heavy hammer to do this. Try not to get the track twisted in the hub or it may jam. If this happens the only solution is to try and crack the bearing using a cold chisel.

In most cases the inner track of the bearing will come away with the outer. But in a few cases the inner track will stay firmly in place on the stub axle. You may be able to get it off using a nammer and drift, in the same way that you removed the bearings from the hub. But on some cars you may end up having to hire a bearing puller.

Your manual may tell you if this is likely. In these cases you will have to consider the cost of hiring tools compared with having the job done professionally — though a bearing puller can come in handy for making a number of repairs easier and may pay for itself in the course of a few jobs.

3. Fitting new oil seal into place on the stub axle

It is not a good idea to leave old bearing parts in place simply because they are hard to get off. The tracks may be imperceptibly worn or oval, in which case the new parts will have only a very limited lifetime and you will end up having to do the whole job properly anyway.

Once you have cleaned all the parts thoroughly in paraffin you can compare the new tracks and bearings with the old parts to make sure they are the same. Save the old tracks because they are very useful for driving in the new ones (**fig 2**) — never hammer directly on to the new bearing tracks.

Pack the hub containing the new tracks with a generous amount of HMP grease and pack the new bearings thoroughly as well. Install a new seal by tapping it home with a hammer and a block of wood (**fig 3**). The assembly can now be refitted to the stub axle. Tighten the large central nut to the recommended torque setting to make sure that the tracks are pushed all the way on to the stub axle and that the bearings are firmly bedded. Now loosen the nut and adjust the bearings as described in Step 2.

MacPherson strut overhaul

If your car sits lower than normal at one corner, or it seems to wallow when you are going round corners, it is time to fit a new spring or a replacement damper to your MacPherson struts

MacPherson struts, like other suspension components are designed to last for tens of thousands of miles. Yet often the coil spring or damper becomes worn or damaged and needs to be renewed.

Replacement springs and dampers are available for practically all types of MacPherson strut. It is then simply a matter of removing the damper from the car so you can renew the damaged parts.

When to do this job
If you car is fitted with MacPherson struts and the springs or dampers show signs of wear or damage

What this job involves
Removing the front strut
Removing the spring and damper
Reassembling the damper and spring
Servicing rear struts

Points to watch
Some MacPherson strut units are available on exchange, but the damper tubes must be free of damage and the stub axle welding must be sound

To do this job
Tools: Spanners; screwdrivers; jack; axle stands; ball joint splitter; spring compressors; a large vice or Workmate would be useful; and perhaps: Mole grips; Allen key; Stillson wrench
Materials: Replacement coil spring; strut; bushes
Time: At least three or four hours for each strut
Degree of difficulty: Straightforward, provided you have the correct tools. But components are heavy, so strength and determination required

If you have the job professionally done . . .
Is the ride comfortable and smooth, particularly when cornering or braking? Are the dampers clean and free from leaks? Does the car return quickly to its normal height if you bounce the wing?

STEP 1 LOCATE THE STRUTS

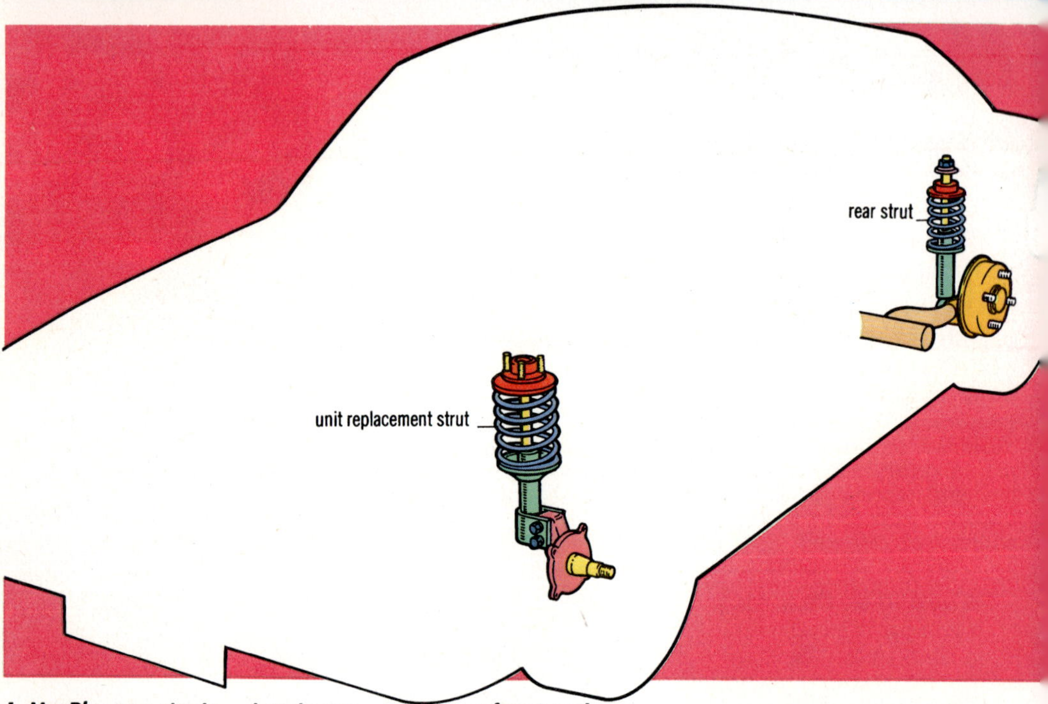

1. MacPherson struts — two types — can be on front and rear

If you are not sure whether your car is fitted with MacPherson struts, look in the car handbook or workshop manual. A quick check on the car will then confirm where they are located (**fig 1**). You are unlikely to have MacPherson struts at the rear of the car and not at the front, so check the front first.

Start by looking under the bonnet. You should be able to see the top end of the strut poking through each inner wing — almost directly above the wheels. MacPherson struts fitted to the rear of the car look very much the same and you will be able to see them by peering behind the rear wheels. If you want a closer look, you are going to have to jack up the car and remove the road wheels. Remember to chock the wheels left on the ground.

There are two types of MacPherson strut — the unit replacement type and the integral strut. The unit replacement strut has a detachable top half which can be removed so that the spring or damper can be renewed. The integral strut, on the other hand, cannot be unbolted in the middle. Instead you will have to remove the whole strut if you want to renew any parts.

Now examine the strut more closely (**fig 2**). The top is surrounded by a large rubber bush and the top mounting. This mounting bolts up into the inner wing from below. Turn the steering wheel to full lock and look under the wheel arch. The MacPherson strut assembly consists of a large single damper surrounded by a coil spring. The lower end will normally be bolted to the track control arm, the steering arm and the hub carrier (**fig 3**). Make a note of whether your car is fitted with the unit replacement strut or the standard strut so you can order the correct parts should you detect any

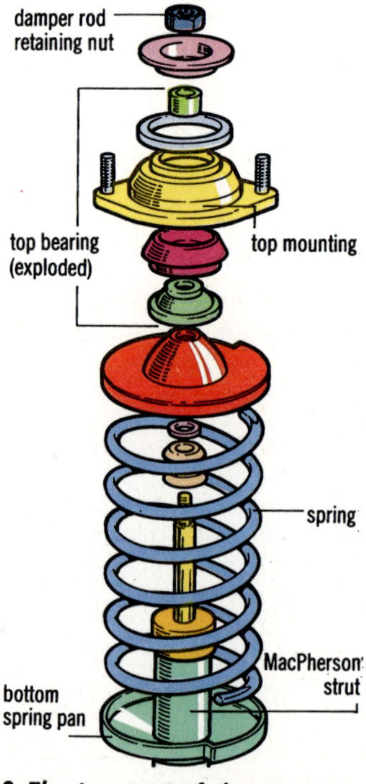

2. The top part of the strut

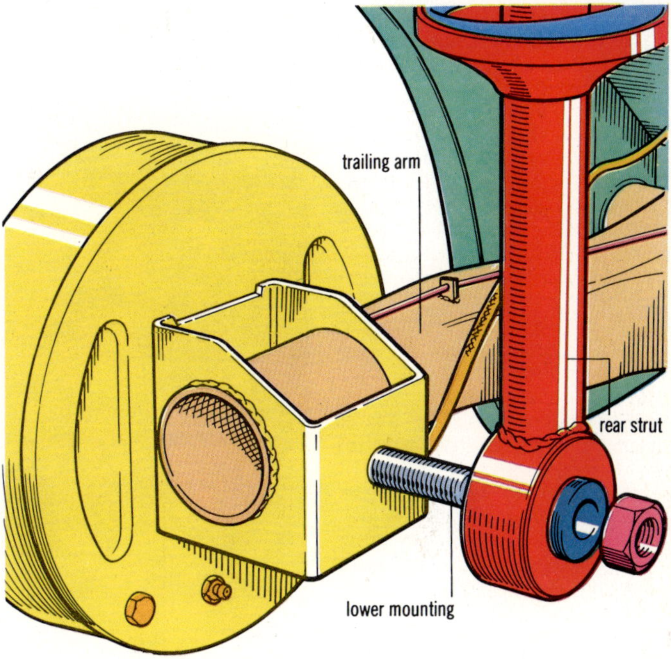

4. The lower end of a rear strut

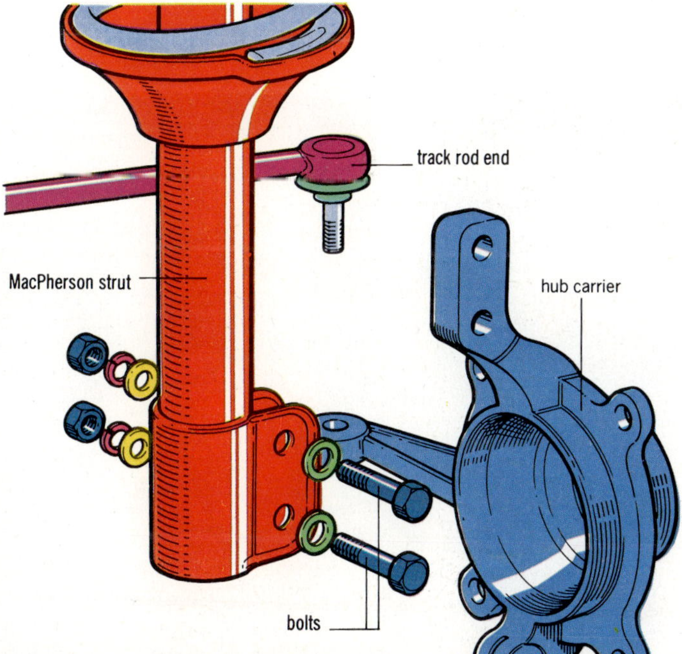

3. The lower end of a front strut

damage or wear.

The rear damper may be the unit replacement type or integral type. If the integral type is fitted, examine the lower mounting carefully. As **fig 4** shows, the damper is usually secured by a single nut. The upper fixings are exactly the same as those fitted to the front dampers. To get to them, open the boot or raise the tailgate and look for the mounting along both inner wings almost directly above the wheels. You may have to remove a trim panel to get at the top fixings — it should just pull off.

As you examine the struts look for signs of wear or damage — for example, misshapen springs or struts which are badly scored or leaking, for example (see pages 184 to 191). You should bear in mind that manufacturers recommend that you renew MacPherson struts in pairs to keep the suspension evenly balanced.

STEP 2 REMOVE STRUT — UNIT TYPE

First jack up the car and place axle stands under the chassis, well clear of the suspension. Point the wheels straight ahead, raise the bonnet and remove the road wheel to gain access to the strut you intend to work on.

Before you start to dismantle the strut it is important to make a note of the exact position of any washers, rubbers and spacers, so you can reassemble them correctly. If the fixing bolts look dirty or rusty you should clean them vigorously with a wire brush, then lubricate the threads with penetrating oil.

Before you start, place a jack directly under the track control arm. Raise the jack so that it takes all the weight of the suspension. Then undo the two bolts that hold the hub assembly on to the bottom of the strut (**fig 2**). Each bolt will have to be driven out. One of them may be an eccentric — used to alter the wheel camber — and will be recognizable by its offset washer or flange. If this is the case, mark the head of the bolt by scratching a straight line across it with a nail, so that it can

1. Using a jack to lift the suspension arm

ALTERNATIVE 2 REMOVE STRUT — INTEGRAL TYPE

First jack up the car and place it on stands. Then disconnect the flexible hose and rigid brake pipes which are joined to each other and to the strut by a union nut plus a lock nut (**fig 1**). Undo both nuts and wrap the ends of each pipe in tape to stop dirt getting in. Tie the flexible hose up out of the way to minimise the amount of brake fluid leaking out.

Undo the nut that holds the track rod end to the steering arm and break the taper joint. The strut is now held in place by the track control arm plus the top mounting bolts. But in order to remove the track control arm, you often have to unbolt the anti-roll bar.

The anti-roll bar is usually

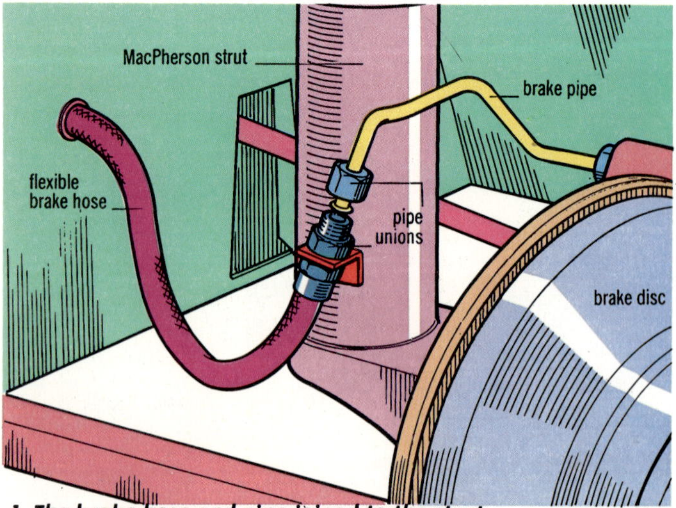

1. The brake hose and pipe joined to the strut

2. Unbolting the hub assembly from the strut

3. Undoing the top mounting bolts

be replaced correctly later.

To release the upper part of the strut, unscrew the two or three nuts holding the strut to the wing (**fig 3**). Do not unscrew the central nut between them (which may be covered by a protective rubber bung) otherwise you will release the coil spring — with disastrous results. Lean under the wing and support the strut as you undo the bolts. Then lift the strut out and take it to a workbench or flat surface so you can service the spring and damper.

At this stage, have a quick look at the strut if you started this job because of signs of fluid seeping from the damper. If it is now clear that the fluid was coming from the damper, then continue the job. But if the fluid is not coming from the damper, suspect a leaking brake pipe and investigate accordingly.

held on by a U-shaped bracket at each side. It is connected to the track control arm by a short link, a clamp or it may pass through a hole in the arm.

Undo the link, the clamp or the large nut behind the track control arm first and then the U-brackets. Then the anti-roll bar will drop away.

Undo the long bolt now that forms the inner pivot of the track control arm — punch the bolt out with a light hammer and a long drift.

Alternatively you can undo the bolt on the bottom of the strut where it fits into the arm. Break this taper joint with a ball joint splitter, then lever the track control arm out of the way.

Then free the upper end of the strut by unscrewing the three nuts or bolts holding the strut to the wing. On no account should you release the nut in the centre of the mounting. Support the strut as the nuts or screws are released. You may need someone to help you do this as the strut is heavy and may drop suddenly. Once the strut is free, lift it out — the wheel hub and brake assembly will still be attached.

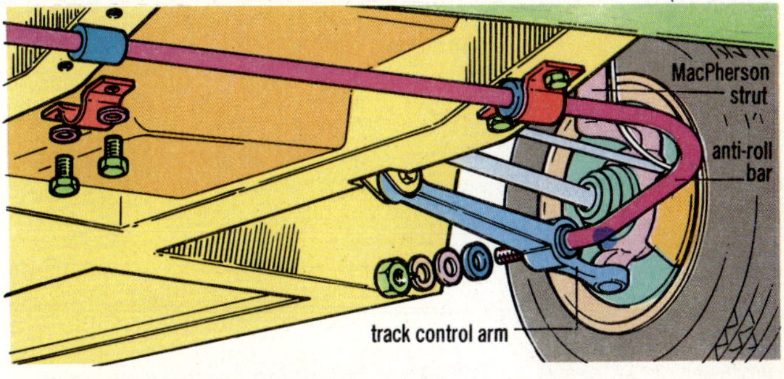

2. The anti-roll bar is held by U-brackets

STEP 3 REMOVE THE SPRING AND DAMPER

1. Freeing off the top mounting

2. Removing the top mounting

Clamp the strut in a vice or a portable work bench such as a Workmate. If you are working on an integral strut it may be impossible to get it into the vice because of its size; if so, lay it upright on your workbench. the strut is filled with oil, so keep the upper end higher than the lower end, to prevent spillage.

To remove the coil spring you need a pair of coil spring compressors. These can be bought from a large car accessory shop or hired. (See Fact file — Spring compressors). Fit the compressors on opposite sides of the spring; make sure that at least four of the coils are clamped. Tighten both compressors evenly until pressure is released from the spring pan fitted to the top of the strut.

The next job is to release the bolt on the damper rod — the central bolt holding the damper in place. The damper rod may be covered by a protective rubber bung; if so, remove this first with a screwdriver.

Before you start, measure the amount of thread showing through the retaining nut. This will allow you to replace the nut in exactly the same position during reassembly. Also make a note of the position of each washer and spacer in the assembly.

The damper rod may have one retaining nut or two. They tend to be difficult to remove so before you start work it is a good idea to give them a few squirts of dismantling fluid to free them (**fig 1**). If only one nut is fitted look to see whether the top end of the damper rod is flattened. If so grasp the flattened end with Mole grips (**fig 2**) and undo the retaining nut with a spanner or socket. With some types of struts you have to insert an Allen key in the top of the damper rod and use this to steady the assembly while you undo the retaining nut. If two nuts are fitted hold the bottom nut with a spanner and then undo the top nut to release the damper.

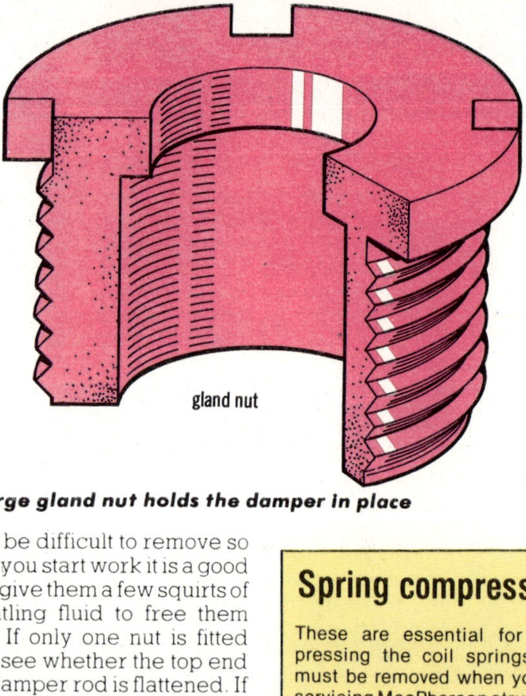

5. A large gland nut holds the damper in place

Spring compressors

These are essential for compressing the coil springs that must be removed when you are servicing MacPherson struts and other dampers. To compress most springs you need a pair of compressors. The compressor consists of a long adjustable threaded bolt with a metal hook at each end. Some types cannot be adjusted; but these are not suitable for most DIY jobs since the springs need to be compressed before you fit the compressors. Do not service a strut without using a compressor.

3. Lifting off the compressed spring

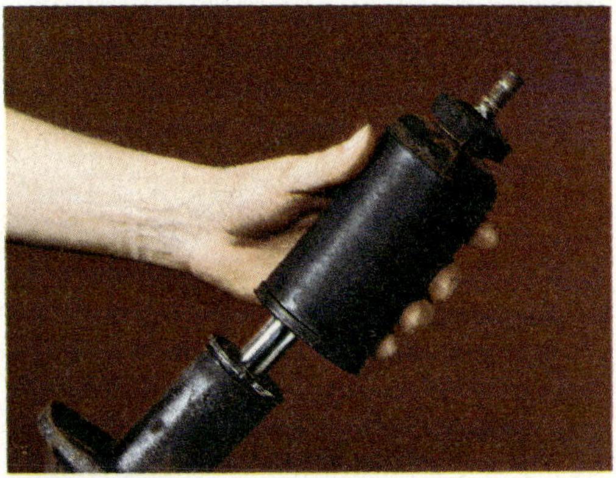
4. Removing the damper gaiter

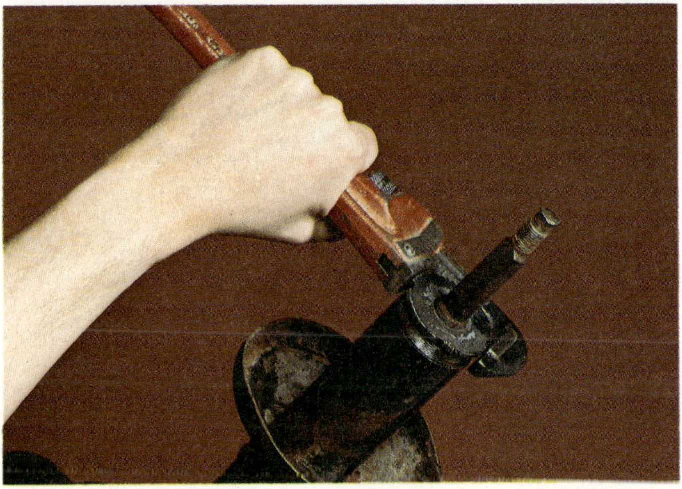
6. Using a wrench to remove the gland nut

Once the damper rod has been released, lift off the top spring pan, the spacers and finally the compressed spring itself (**fig 3**). Some dampers are fitted with a rubber bumper; if so pull this off too.

If the damper is damaged or faulty, you can remove it next. There may be a protective rubber gaiter around the middle of the damper (**fig 4**). If so, remove this first; it is sometimes held by a Jubilee clip secured by a bolt or screw.

The damper is held in place by a larger gland nut (**fig 5**). You may be able to remove this with a larger spanner or a Stillson type wrench (**fig 6**) or you can clamp the nut in the vice and turn the strut anticlockwise. Other nuts have two or three slots and need to be removed with a special spanner. You may be able to remove this type of nut by turning the strut on its side, and knocking the nut round with a hammer and large screwdriver.

If the damper has never been serviced, the piston rod will just pull out of the strut along with the piston and a few smaller components. Turn the strut upside down and let the oil drain out into a bowl or on to a pad of newspaper.

Then clean the inside of the damper tube with petrol-soaked rag on the end of a stick.

FACT FILE

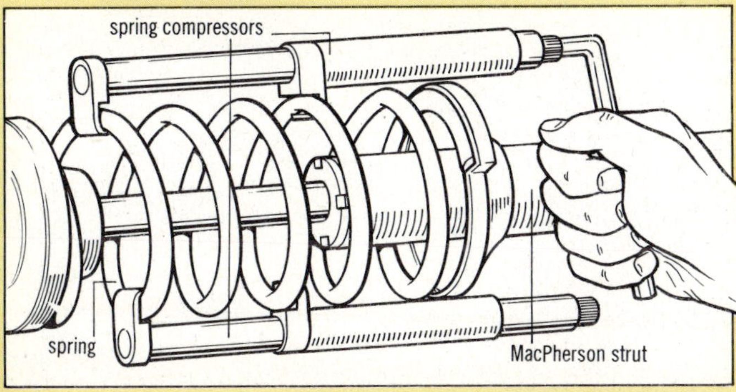

STEP 4 REASSEMBLE DAMPER AND SPRING

In most cases you can buy a sealed cartridge that fits inside the old damper tube and replaces all the separate bits of the old damper. Where no cartridge is available, you will have to buy a complete new strut though strangely enough, there is not much price difference between the two.

When you are fitting a cartridge, pour a little engine oil into the damper tube and shake it up to prevent the tube rusting from the inside.

Once the damper is in position, retighten the gland nut and check that the damper is held securely by trying to move it backwards and forwards with both hands. Then replace the rubber bumper, if fitted.

Fit the coil spring next. If you are renewing the spring, compress it first as described in Step 3. You should first ensure that they are the same rating as the old springs. The new springs must be compressed to the same compression rate as the old ones. The best way to do this is to clip the same number of adjacent coils together. Also the clips should be applied in the same position on the new springs as the old springs.

To fit the spring, check that it is the right way up and then locate the bottom end in the lower spring pan. This is shaped to fit the spring exactly. Then re-fit the top pan, the rubber bumper, the top mounting and finally the damper rod nut.

The rubber part of the top mounting must be in good condition as it forms the only steering bearing. The hole in the mounting is a very tight fit on the damper rod, so oil the bore before you fit it and tap it into place.

Then screw the retaining nut(s) on to the damper rod and tighten it up until the required amount of thread pokes through. If the nut is covered by a protective rubber bung, replace this as well.

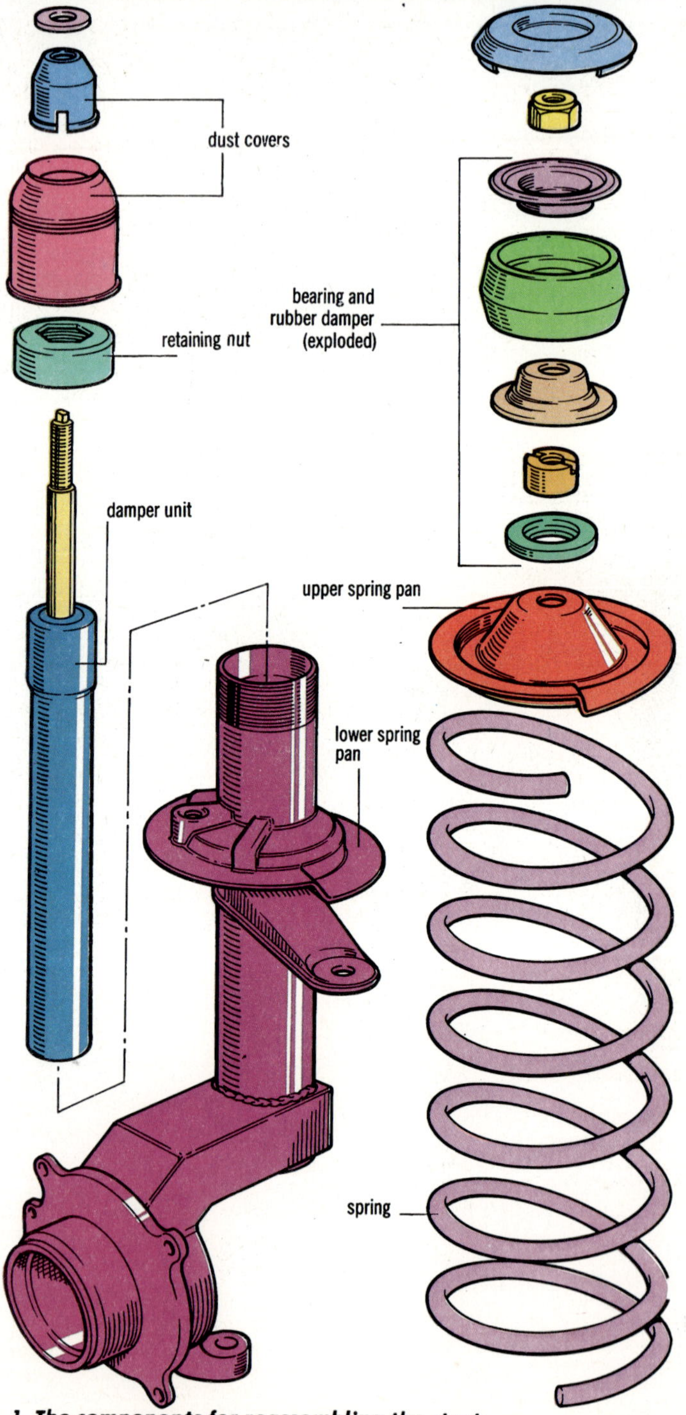

1. The components for reassembling the strut

STEP 5 REFIT THE FRONT STRUT

1. Applying oil to top mounting

2. Using a wrench to replace the track rod

3. Offering up the strut

Before you refit the strut it is a good idea to clean all the dirt out of the inner wing with a wire brush, particularly around the mounting point.

Offer up the strut. Insert it first into the upper mounting and then push the lower half across until the fixing holes are aligned correctly (**fig 3**).

If you are fitting a unit replacement strut, insert the lower fixing bolts and tighten them fully. If the lower nut is an eccentric, make sure that it is returned to exactly the same position as before using the line you have marked on the bolt head as a guide.

When fitting an integral strut, first rebolt the strut to the suspension arm. Then move the track rod arm into position. It is a good idea to use a Stillson type wrench to hold the arm as you do this. Then by juggling with the wrench you can slot the bolt easily into place (**fig 2**). Press the tapered joint into place and refit the retaining nut and split pin, if necessary. If you removed the anti-roll bar, refit this next. Push the bar into place, fit the U-brackets over the top and retighten the bolts or screws.

With the damper rod assembly in position you are now ready to decompress the spring. Release each coil evenly, a small amount at a time, so that the spring returns to its original position. Then, taking great care, remove the spring compressors.

Next retighten the nuts on the top mounting. If your strut is the unit replacement type, lower and remove the jack which you placed under the lower suspension arm.

To check that the strut is fitted correctly, grasp hold of the strut at either end and try to move it back and forth. There may be just a little free play, but the strut should essentially feel firm and stable.

Refit the two brake pipes using the union nut; you will have to top up the brake master cylinder and bleed the brakes with the help of an assistant before it is safe to drive off. Replace the wheel and lower the car to the ground.

Finally, tighten the damper rod nut again. This nut must be tightened to a specified torque as listed in your workshop manual.

STEP 6 SERVICE THE REAR STRUT

Removing the rear struts and fitting new ones is very much the same on the rear as it is on the front. In most cases a coil spring is used in just the same way as on a front strut but many Fiats are fitted with a long leaf spring that serves both sides of the car. Here the only difference is that you must use a scissor jack under the track control arm to control the spring. It is not usual to find an anti-roll bar at the rear.

Rear wheel bearing service

Rear wheel bearings on a live-axled car need attention from time to time. Fitting them is a garage job but you can still do most of the work yourself

All cars have bearings which allow the rear wheels to spin, but the way the bearings do their job varies from car to car. This article deals with live axle rear wheel bearings fitted to rear wheel drive cars.

Rear wheel bearings are generally reliable components but if they do give trouble you should get them seen to or do the job yourself. With the types of live axle most commonly found on modern cars, the actual job of getting the old bearings off and the new bearings on is usually best left to a garage with specialist equipment, but you can still significantly reduce the bill by doing the preparatory and reassembling work yourself.

When to do this job
As recommended by the manufacturer or if you hear rumbling noises from rear wheels

What this job involves
Checking the source of the noise
Removing the halfshaft and bearing
Reassembling rear axle

Points to watch
Sometimes the oil seal for the bearing is separate but on most modern cars, the oil seal is part of the bearing itself. Either way, take great care not to damage it

To do this job
Tools: Spanners; hammer; pullers (maybe); slide hammer (maybe); jack; axle stands
Materials: New bearings; axle oil
Time: Depends on difficulties but allow a whole day
Degree of difficulty: Removing halfshaft is straightforward but fitting new bearings is usually a garage job

If you have the job professionally done . . .
Does car run without undue noise from rear axle?
Does any oil leak from rear axle?

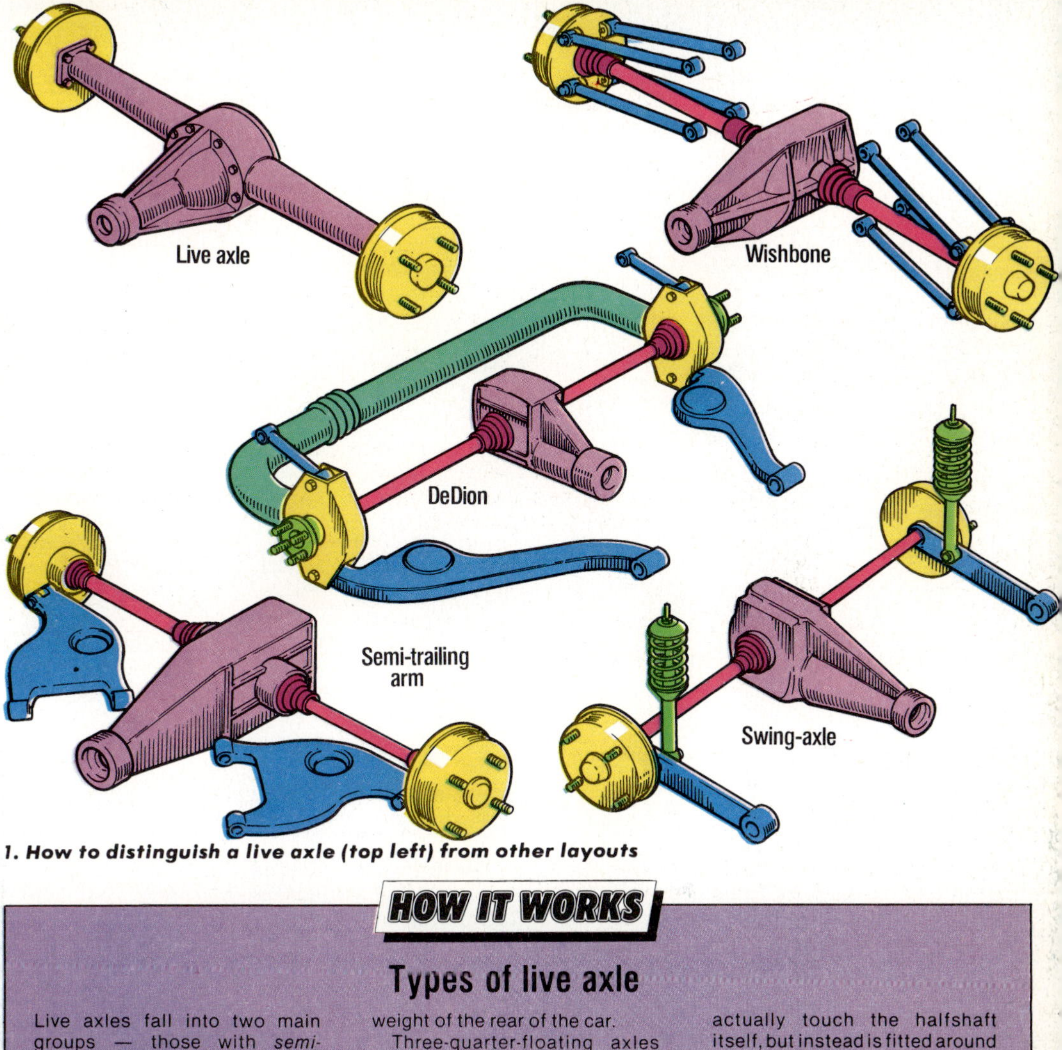

1. How to distinguish a live axle (top left) from other layouts

HOW IT WORKS

Types of live axle

Live axles fall into two main groups — those with *semi-floating* wheel bearings and those with *three-quarter-floating* wheel bearings. You will find that the procedures for working on the two types are rather different. Check with your manual or dealer to find out which type you have as they look similar from the outside.

The main difference between the two is the way in which the halfshafts are supported at the wheel ends of the axle. In the more common semi-floating type the bearing fits around the outer end of the halfshaft inside the axle casing. With this arrangement the bearing and the halfshaft support the entire weight of the rear of the car.

Three-quarter-floating axles are less common but are still found on many cars, especially older designs. With this type of axle the bearing does not actually touch the halfshaft itself, but instead is fitted around the outside of the axle casing inside a hub, so the halfshaft does not support the weight of the car.

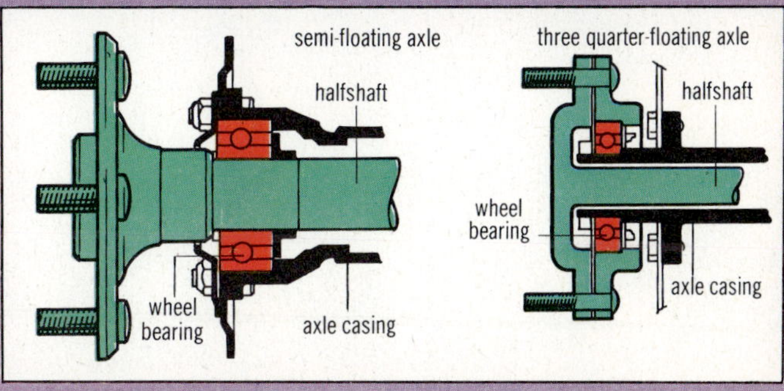

STEP 1 FIND THE FAULT

Unlike the front wheel bearings which are lubricated by grease, bearings fitted to most rear wheel drive cars are lubricated by the oil inside the axle casing — usually a thick SAE 90 hypoid oil. Problems often arise with the bearings when this oil level is allowed to run low for a period — perhaps because of a leaking pinion seal in the differential casing. The result is that the bearings are starved of oil and quickly wear out. It is usually fairly easy to tell when this has happened because there will be an obvious grinding or rumbling sound coming from the back axle.

However, before you decide to replace the wheel bearings it is worth checking that the noise is not coming from the differential gears or bearings, or even the tyres. Jack up the rear of the car so that the wheels clear the ground and support it securely with axle stands — make sure that both front wheels are chocked securely. Make absolutely certain that the car is solidly supported, then release the handbrake and check that the gearbox is in neutral.

Begin by making a visual inspection of the wheels and tyres, spinning them by hand to see whether they turn smoothly

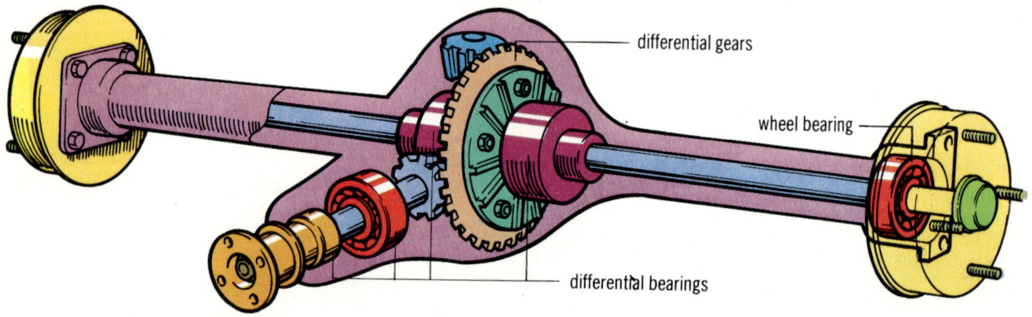

2. The parts of a live axle that may become noisy

STEP 2 PREPARE FOR THE JOB

Once you are certain that it is the wheel bearings that are at fault you will have to find replacements as well as any new gaskets and oil seals which may be needed. However, before buying the parts you should check with your dealer or your manual to find out if there are any special tools needed for the job and how much you will have to pay for them. You may need a torque wrench and you are also likely to find the job much easier if you have a special slide hammer for pulling out the halfshaft or hub — in some cases this tool is essential. In the case of three quarter-floating types you may need a variety of pullers to do the job. Usually these tools can be hired but you may find the cost of hire plus the new parts almost equals the amount a garage would charge.

The next point to check concerns the removal of the bearings. With most types a large hydraulic press has to be used to remove the old bearing from the halfshaft and to refit the new one. Sometimes this job has to be done after the bearing has been heated up. With other types special bearing pullers are needed to pull off the bearings or to extract them from their housings. This stage of the work is best left to your dealer, local garage or even a specialist so you will have to make arrangements beforehand to have the job done on the day

1. Loosening the differential drain plug

1. Listening for wear using a wooden rod as a stethoscope

and do not wobble as they spin. Check the tyre treads and sidewalls carefully for signs of damage. Make sure the tyres are the right size and type for the car — worn, damaged or incorrect tyres can make excessive noise at times and it is possible that you have mistaken this for bearing noise. Bear in mind that if you fit extra-wide tyres on the back of the car you will put more load on the rear wheel bearings possibly causing them to wear out prematurely.

Next, ask someone to sit in the car, start the engine and engage first gear while you listen for noises around the rear wheels. Take care not to get too close to the wheels as they spin round. Sometimes the bearings will only make noises at higher speeds so your assistant may have to change to a higher gear. Usually it will be clear whether or not the noise is coming from the outer ends of the axle close to the wheels, or from the middle close to the differential.

If you are still not certain about the precise source of the noise you can use a long wood or metal rod as a stethoscope to pinpoint the sound. Touch one end against the axle casing, first in the middle and then at the two ends, while you hold the other end of the rod to your ear. This should amplify the sound enough for you to tell where the noise is coming from.

If you are absolutely sure that the noise is from one or both of the wheel bearings you can go ahead and start work on the axle. If the sound seems to be coming from the differential you should go to a dealer or a specialist to have your diagnosis confirmed. You may end up having to fit an entire reconditioned axle assembly, although a noisy differential will often carry on for years without giving further trouble.

you plan to do the repair. You may find that the garage is more willing to do the work if you buy the parts from them as well. Even though you will have to pay for this service you could still save money by doing the rest of the work yourself but the saving may not be worthwhile.

When you are ready to begin, bear in mind that if you do not jack up both sides of the car evenly you will have to drain the differential oil to stop it running out of the end of the axle when you remove the halfshaft and bearing. Use a spanner or a drain plug tool to loosen and remove the filler plug and then the drain plug at the base of the differential casing (**fig 1**).

The bearings should be removed from both sides, but if you do not want to drain the oil from the axle you can jack up one side of the car higher than the other so the oil runs away from the higher side.

Now loosen the nuts or bolts holding the road wheels in place and jack up the car and support it securely.

FACT FILE

Slide hammers

Some components, like halfshafts and engine bearing retainers, need a jolt to free them from their housings. Often the only way to achieve this is by using a slide hammer which gives a sharp pull in the right direction.

The slide hammer consists of a heavy weight which slides along a bar. The bar is bolted to the component which needs to be freed. When the weight hits the end stop it jars the component and shocks it free.

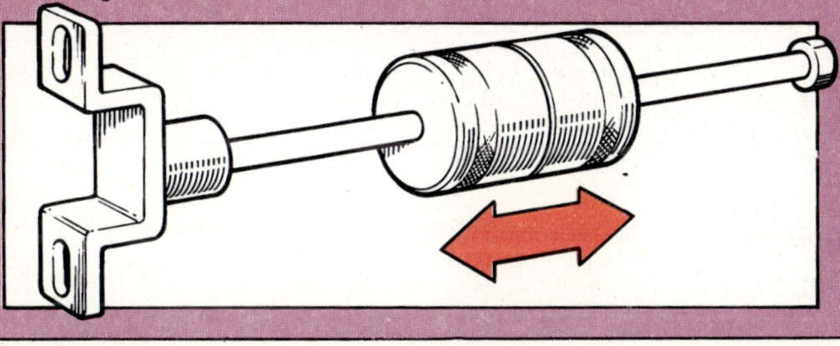

STEP 3 REMOVE BRAKE ASSEMBLY

If your axle is a semi-floating type, you may first have to take off the entire brake assembly to gain access to the bearing assembly.

First, disconnect the handbrake cable from behind the brake backing plate. Usually the cable is attached to the handbrake lever by a clevis pin — this is locked in place by a split pin so use a pair of pliers or a small screwdriver to remove it before sliding out the clevis pin (**fig 1**).

No matter which type of axle you have, the next step is to

1. Disconnecting handbrake cable from brake backplate

STEP 4 DISMANTLE SEMI-FLOATING AXLE

1. Using socket to unbolt retainer plate

2. Retainer plate unbolted from axle flange

On semi-floating axles the bearing is fitted around the halfshaft so these two components have to be removed together. If you look behind the halfshaft flange towards the end of the axle casing you will probably be able to see four nuts or bolts which hold the bearing retainer plate to the axle casing. Sometimes these nuts or bolts are on the other side of the flange so that they are facing you once the brake linings have been removed — they often hold the brake backing plate in place as well. Undo the bolts a little at a time with a spanner, working diagonally from one bolt to the one opposite to avoid distorting the bearing cover. Sometimes, when the bolts are threaded in from the front, there will be an access hole cut in the halfshaft flange so that you can use a socket spanner (**fig 1**).

At this point, in many cases, you will have to disconnect the hydraulic brake pipe from the back of the wheel cylinder and plug it to minimize the loss of brake fluid. Sometimes the back plate can be left hanging on the end of the axle casing and the hydraulic connection does not have to be disturbed. The backing plate and bearing retainer plate should now be free from the flange (**fig 2**) — make a note of the shims and gaskets behind it.

A few cars do not have bolts which hold the bearing in the axle casing. Instead a large internal circlip is fitted. Use a pair of circlip pliers to release it from its groove in the casing.

Next, the halfshaft and the wheel bearing have to be pulled out of the axle casing. This is where you will probably need the slide hammer. After a few fairly sharp blows with the hammer, the bearing should start to leave the casing so that the shaft and bearing can be slid out of the casing (**fig 4**).

In many cases it is possible to do this without a slide hammer, although the job is more difficult this way. Begin by refitting the road wheel to the flange and

2. Bearing retainer plate bolted through backplate

remove the brake drum. Sometimes this is held in place by one or two countersunk screws, sometimes by bolts or nuts and sometimes just by the wheel nuts or bolts.

Once the drum is off you will be able to see the flange of the halfshaft or the hub. With three quarter-floating axles, the rest of the brake components can be left in place, while with semi-floating types you will have to remove the brake shoes and springs as they will almost certainly get in the way of the bearing retainer plate.

3. Fastening slide hammer in place on halfshaft flange

4. Drawing bearing out of axle casing with slide hammer

tighten the nuts or bolts enough to hold the wheel firmly in place. Now get under the car, behind the wheel, and hold a stout block of wood against the inside edge of the rim.

Hit the block of wood with a heavy hammer and then turn the wheel rim 180° and repeat the process. Several blows should push the bearing out of the casing — never hammer directly on to the wheel rim otherwise you will damage the wheel or possibly the tyre.

Handle the halfshaft with care as you slide it out of the casing and take it to a garage to have the bearing removed — dropping it on a hard surface is likely to damage the splined tip or even distort the whole shaft.

If you examine the shaft you may notice that the bearing is held in place by a strong circlip. However, it is also pressed into place with great force and can only be removed properly by using a heavy-duty hydraulic press — do not try chiselling the bearing off the shaft as the hardened steel may shatter and injure you. Take the shaft to your local garage or machine shop where they will press the bearing on for a couple of pounds. Remember to take along the bearing retainer.

STEP 5 REASSEMBLE THE AXLE

Once you have had the new bearings pressed into place on the halfshafts lubricate them with the same type of oil used in the axle — usually hypoid SAE 90. Do not try to prise away any oil seals from the bearing. Make sure that the splines and the rest of the shaft are spotlessly clean before sliding the assembly back into the casing. You may have to twist or rock the shaft a little as you insert it to make sure that the splines engage with the differential unit in the middle of the axle — you should never have to force the

1. Easing halfshaft into place so that splines engage

ALTERNATIVE 4 DISMANTLE 3/4 FLOATING AXLE

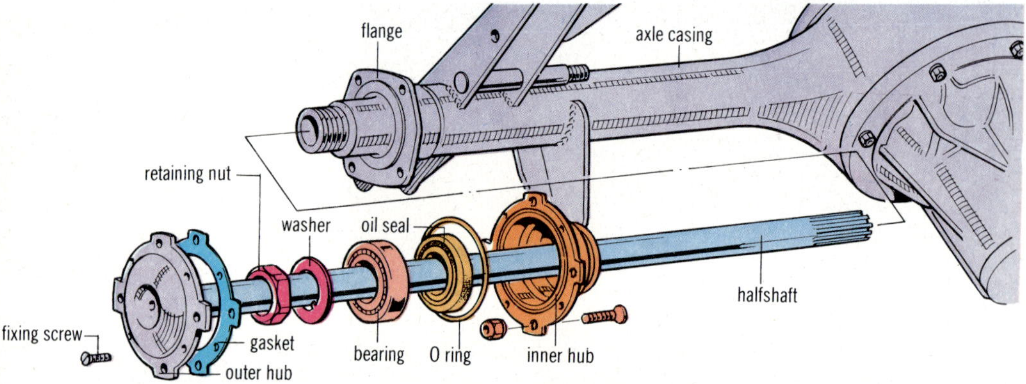

1. The components of a three-quarter-floating axle with two-piece hub

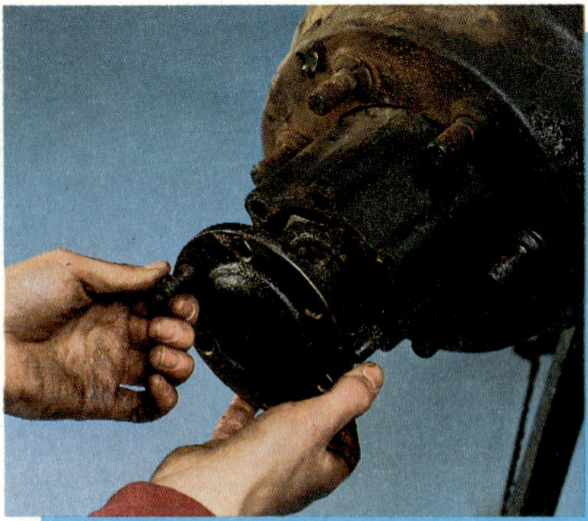

2. Unbolting halfshaft flange from hub

3. Loosening central nut with screwdriver

halfshaft into place except for when the bearing is starting to enter the casing. Once this has started to happen you should be able to tap the end of the half-shaft all the way home using a hammer and a piece of wood (**fig 2**) — do not strike the shaft directly with the hammer, or you may damage it. Check that the bearing is going into the casing squarely as you are tapping it home. Finally, tighten the flange bolts according to the recommendations in your manual or fit the circlip back into its groove in the axle casing.

2. Tapping halfshaft home with hammer and block of wood

The procedure for removing the bearing from a three-quarter-floating axle is rather different. Once you have taken off the brake drum the rest of the brake assembly can be left intact. First, undo the screw or screws which hold the halfshaft flange to the bearing hub — since there is no bearing holding the shaft in the axle casing the shaft can be pulled out and placed to one side. Note that there is usually a paper gasket and a rubber O-ring fitted between the inside of the halfshaft flange and the bearing hub. A new gasket and O-ring should be fitted on reassembly.

Now you will be able to see the large nut which holds the bearing hub to the axle. Use a screwdriver or a small chisel to bend back any tabs which lock the nut in place and then use a large spanner to undo the nut. You may have to borrow a spanner large enough for the job or tap the nut round with a screwdriver (**fig 3**). With the nut removed, the hub can be pulled away using a slide hammer.

You do not need a hydraulic press to remove the bearing (**fig 4**). When the hub assembly is removed, place it with its outer face downwards on a piece of wood and use a hammer and a soft metal drift to tap the bearing out of the hub (**fig 5**). Work your way around the edge of the bearing so that it is forced out evenly. Use a screwdriver to prise out the oil seal, making a note of the way the new one should be fitted. Reassembly is a straightforward reversal of the dismantling procedure, but make sure the new bearing is well coated with oil before tapping it back into the hub.

4. Removing outer bearing from hub

5. Drifting out inner bearing assembly

Replacing half-shafts

Replacing a broken halfshaft is a tough job that needs careful planning. Hire special tools in advance and have an assistant on hand

Live rear axles are extremely simple, robust and reliable components but when they do fail it is generally something major. Sudden loss of all drive even if the propshaft is still turning, accompanied by scraping or grinding noises from one side of the axle, is a sure sign that a halfshaft has gone. Replacing a halfshaft is not particularly difficult, but it can be hard work. If the half-shaft is broken at one end it can be fiddly getting the parts out.

Depending on the age and value of your car, fitting a guaranteed secondhand axle from a breaker's yard might be the best solution. But if you do decide to replace just the half-shaft plan the job carefully. Hire any special tools in advance and try to have someone on hand to help you with the heavy work.

Axle stands are a must if you have to take the differential out — resting the car on ramps will not give enough height.

When to do this job
When you break a halfshaft

What this job involves
Removing broken halfshafts
Removing and refitting differential (maybe)

Points to watch
If you are fitting a new half shaft, always fit a new bearing at the same time

To do this job
Tools: Jack; axle stands; sockets; and possibly: drain plug spanner; slide hammer; universal puller; self-locking wrench; long thin rod; magnet
Materials: Replacement halfshaft; new gaskets and oil seals; Woodruff key (maybe); oil; stiff wire
Time: Allow at least one day and, ideally, a weekend
Degree of difficulty: A tough, awkward job

If you have the job professionally done . . .
Is the transmission smooth and quiet? Has the differential been refilled with oil, if drained? Do the brakes work normally?

STEP 1 TAKE OFF BEARING/SHAFT RETAINER

Most cars have the halfshafts made in one piece with the hub flange, held into the axle by a press-fit bearing, itself retained by a plate bolted over the end of the axle tube. However, other types of shaft are separate from the hub. You can identify these types of shaft by the screws, bolts or single nut in the centre of the hub.

Whichever type you have, start the job by draining the oil out of the axle. This is best done after you have run the car for a couple of miles to warm the oil. Although it is unlikely that oil will run down the axle tubes to the brakes, draining is a sensible precaution. Normally there is a plug (**fig 1**) underneath the axle which has to be undone with a special square headed tool. Sometimes you can improvise with an Allen key or a square bar of steel but it is easiest to buy a special wrench or socket adaptor, though you must check that it will fit your car. If you are lucky, though, your car's drain plug will have a conventional bolt head which you can undo with a socket or ring spanner.

Unfortunately, manufacturers sometimes do not fit a drain plug and you will have to loosen the ring of differential mounting bolts at the front of the axle (see Step 4). Raise the front of the differential slightly so that the oil can drain out of the small gap which will form between the differential and the axle casing. Catch the oil in a suitable container of at least one litre capacity. As the oil is thick it will take a long time to drain completely.

If you cannot reach or see all of the nuts, screws or bolts which retain the halfshaft then you will have to remove the brakes. Disconnect the hand brake cable by taking out the clevis pin. Remove the brake drum and use a pair of heavy pliers to take off the steady springs on the brake shoes.

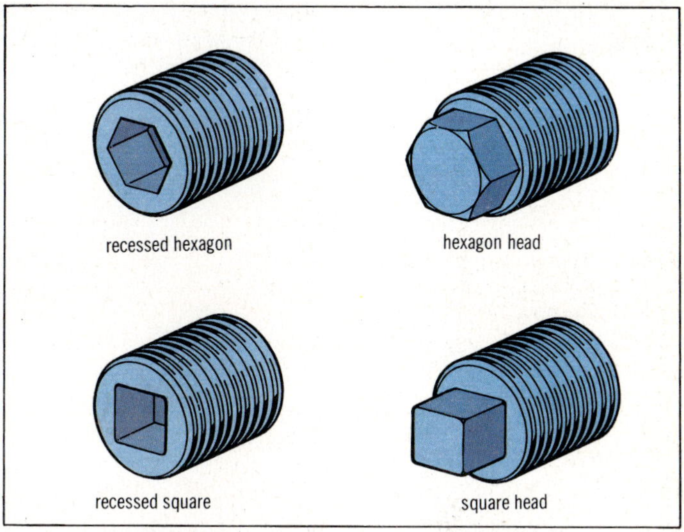

1. The four most common types of axle drain plug

Remove the brake shoes themselves, the pull-off springs and the wheel cylinder. This will give you access to the back plate, the halfshaft and the bearing retainer. Clean off any oil that has leaked onto the back plate.

If you have the more common one-piece halfshaft and hub, turn the hub flange until the hole in it lines up with one of the bolts that hold the bearing retainer in position. Undo the bolt with a socket spanner fitted with a short extension and move on to the next one and so on until they have all been removed.

The bearing retainer is now free to come out of the axle casing.

Some cars are fitted with a circlip instead of a plate to hold the bearing in place. You will need a pair of circlip pliers long enough to reach the ears of the circlip inside the axle tube (**fig 2**). Fit the prongs of the pliers into the holes in the circlip, then squeeze the pliers and pull the circlip out of its groove. Slip a small screwdriver behind the circlip to help you lever it out of the end of the axle casing. The next step is to pull the halfshaft out of the axle tube.

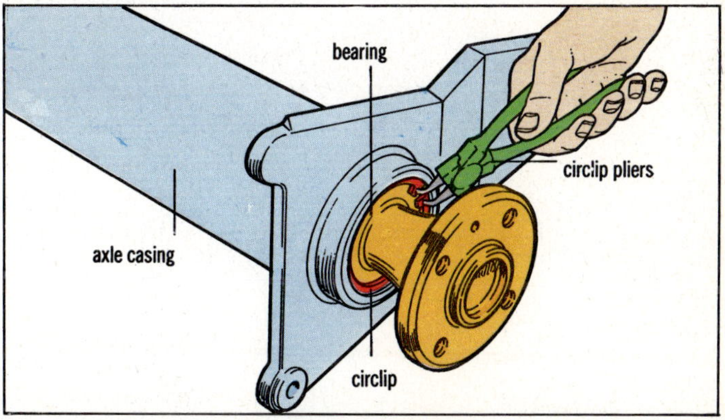

2. Some cars have the bearing retained by a circlip

STEP 2 PULL OUT HALFSHAFT

Once you have removed the bearing retainer, all that is usually holding the halfshaft is the bearing which is a tight fit in the axle tube. If the shaft will not come out — and it probably will not — the best way to remove it is to hire a slide hammer. If you cannot hire one, try bolting the spare wheel into place on the halfshaft flange (**fig 1**) so that you can get a better grip on the shaft.

Do the wheel nuts or bolts up by only a few turns. Hold each side of the wheel and pull it towards you as evenly and as hard as you can (**fig 2**). In effect you are using the weight of the wheel just as you would use a slide hammer. If the shaft still will not move you will have to apply more force.

Do not try hammering the inside of the wheel rim as this

1. If the halfshaft will not pull out, refit the wheel

ALTERNATIVE 2 PULL OUT HALFSHAFTS (OLDER TYPE)

Some older British cars are fitted with halfshafts that do not carry the bearing directly. Instead the bearing is carried in the hub and the flange of the halfshaft mates with the face of the hub. To remove this type of halfshaft, first undo the countersunk screw that holds the halfshaft in place (**fig 1**), and tap the centre of the flange with a hammer. This will separate the two flanges slightly so that you can push a screwdriver in between the flanges and lever them apart. If tapping the flange does not work, fit wheelnuts to two studs diagonally opposite each other and hit those lightly instead.

All that you need to do now is to draw the shaft out of the axle. You will probably find either a paper gasket or a rubber 'O' ring fitted between the halfshaft flange and the hub — make sure that you always fit a new one when you put the axle back together.

Marinas are fitted with an axle design quite different from that fitted to most cars (**fig 2**). To remove the halfshafts the first step is to take off the large nut and washer which is fitted in the centre of the hub. Then remove

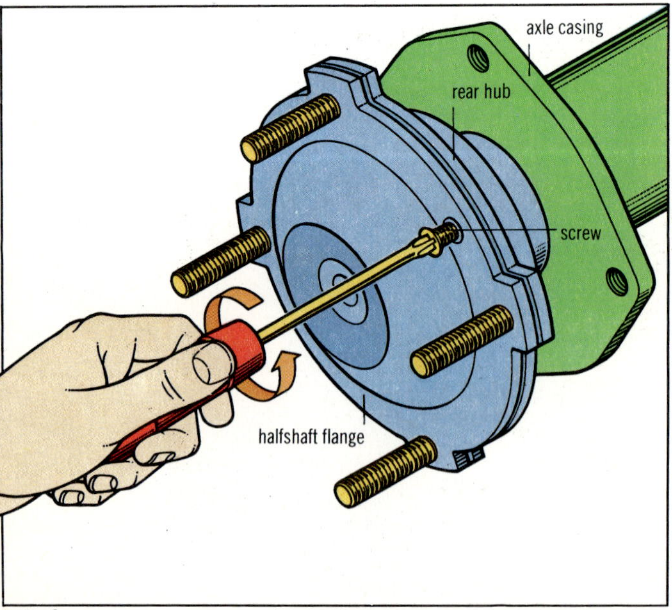

1. Before you can remove the halfshaft, you must undo the screw

2. Use the wheel as a makeshift slide hammer

will damage it. Instead, do the wheel nuts up tight and check once again that the car is firmly and safely supported.

Position yourself under the car so that you can rest your feet on the inside of the wheel. Try to sit so that your legs are slightly bent. Now kick out as strongly as you can. If kicking does not work, arrange some sort of back rest that you can lean against and then use your feet to apply a steady sustained pressure on the shaft.

This method will work in almost every case but if it fails, get a wheel from a scrapyard and bolt it into place. You can now use a sledgehammer on the inside of the wheel without fear of damaging anything important, although take care not to hit the backplate or brake pipes.

the brake drum and the hub. You will probably have to use a universal puller to take off the hub if tapping it all round from behind does not work. Next, disconnect the brake pipe and the handbrake cable — plug the end of the brake pipe to reduce the amount of brake fluid that leaks away, or stretch a piece of polythene over the filler for the brake fluid reservoir and screw the cap over it.

You must now locate the oil catcher which is bolted to the brake backplate by the same four bolts which also fasten the backplate and oil seal retainer. Mark the oil catcher so that you can reassemble it in exactly the same place on the backplate. Undo the four bolts and lift off the oil catcher and the complete backplate.

Behind the backplate you will find another plate and a paper gasket between the plate and the axle. Remove these then turn the end of the halfshaft so that you have access to the Woodruff key. Pull the Woodruff key out of its slot with a pair of pliers then pull the halfshaft out of the axle. If it is tight, refit the large nut and use a slide hammer.

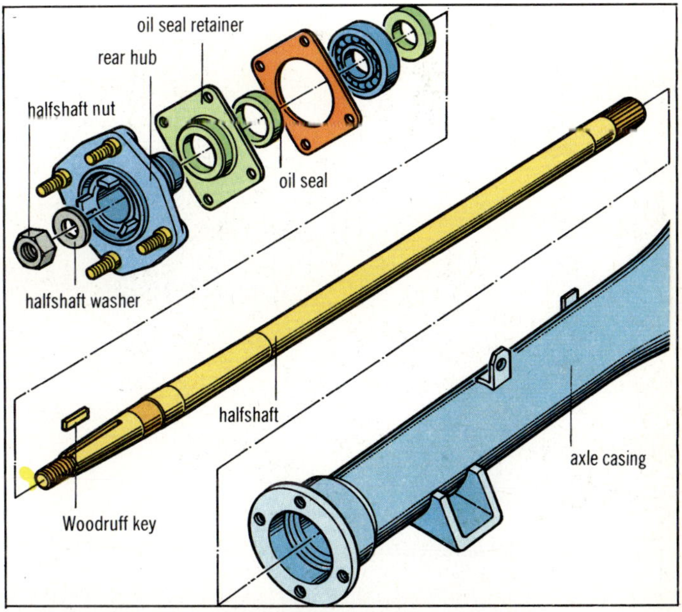

2. Exploded view of Marina type halfshaft arrangement

STEP 3 REMOVE A BROKEN HALFSHAFT

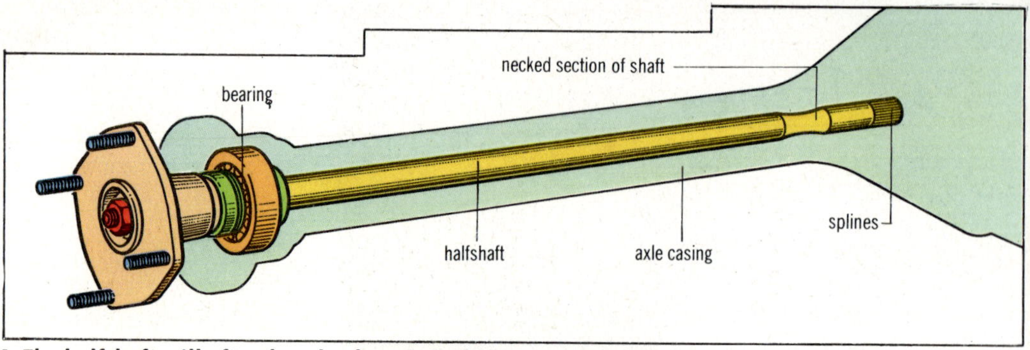

1. The halfshaft will often break where it is thinnest — usually where it is machined

When a halfshaft breaks, it tends to do so at the differential end where the shaft is necked, or made thinner, (**fig 1**). The other place you might find a break occurring is near the wheel where the shaft is machined to accept the bearing.

First remove as much of the halfshaft as you can as described in Step 2. Then inspect the broken end of the halfshaft if you can see it.

If only a short stump of the shaft came out, you should be able to see the rest of the shaft in the end of the axle tube. The splined end of the shaft is a tight fit in the side gears of the differential, so grip the end of the shaft with a self-locking wrench or a large pair of pliers and move it gently up and down, pulling the shaft at the same time (**fig 2**) to free the splines.

2. Removing broken piece

If the break is at the other end of the shaft, near the differential, the stub is usually held quite firmly in the side gear of the differential. There are three methods of getting the stub out but two require you to remove the other halfshaft first (see Step 2).

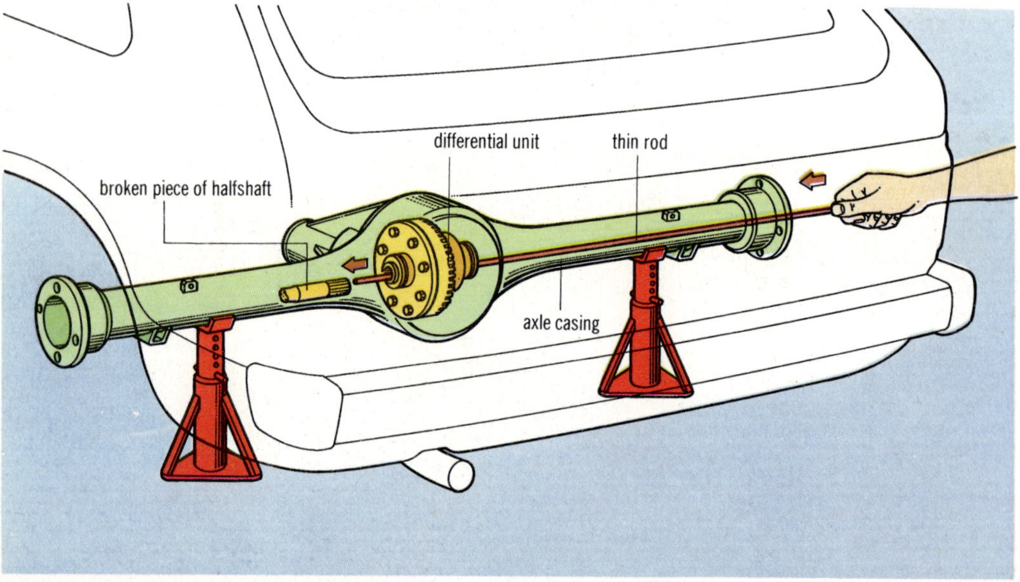

3. If you are lucky you will be able to push out the broken piece with a thin rod

Find a long, thin rod that will reach right across the axle. The idea is to use this from the other side to push the broken stub out. Unfortunately the design of the differential often makes it very difficult to pass the rod straight through into the opposite side gear as the pinion pin gets in the way (**fig 3**). So push the rod through the middle of the axle and when it meets an obstruction waggle it around. Try to support the rod so that it is central in the axle tube and goes through the hole in the side gear, beside the pinion pin and meets up with the stub in the other side gear. As the rod will already have gone through one side gear it will be roughly lined up with the other one. Provided that the rod is thin enough and you hold it absolutely horizontal, you should now be able to knock the stub out of the differential into the opposite axle tube (**fig 3**). Get a long piece of stiff wire and put a right angled bend in it half an inch from one end. You can now use this to get the stub out of the axle tube from the other side by using the bent end as a rake (**fig 4**).

If this method does not work, the other way of dealing with a very short halfshaft stub stuck in the differential is to remove the differential (see Step 4). But this is only possible when the stub is short enough.

Occasionally, the halfshaft breaks in the middle so that you cannot reach the halfshaft from the end of the axle tube, or remove the differential either.

If this happens, find a long piece of thick, soft wire and twist it into a loop at one end. Make the loop about a quarter of an inch larger in diameter than the halfshaft. Now pass the loop down the axle tube and try to lasso the end of the shaft. Push it on to the shaft as far as you can and pull the wire gently. With any luck, it should grip the shaft and draw it out. If it does not and the wire just slips off the end, reduce the size of the loop and try again. If you still cannot get the broken piece of halfshaft out and you have a differential cover on the back of the axle, remove this (**fig 5**). You will be able to see the differential and the side gears, and also where the halfshafts are splined into the side gears. You should be able to prod the broken piece of shaft out of the side gear with a drift (**fig 6**).

4. Rake out any loose pieces with stiff wire or flexible rod

5. Unbolting the rear cover of the differential

6. With the cover off you will be able to push out the stub

STEP 4 TAKE OUT THE DIFFERENTIAL

There are two types of differential that you are likely to come across. The more common type is built up as a separate unit and is bolted into the welded axle case with a ring of bolts at the front. The other type has a cover bolted to the back of the axle casing which you can remove to expose the differential without having to take it out, as described in Step 3.

Start work by marking the flanges of the propeller shaft and the differential so that you can replace the parts just as they were (**fig 1**). This is very important because the prop shaft is balanced carefully to stop it vibrating at high speed.

Now undo the four or more bolts that hold the two flanges together (**fig 2**). Sometimes you also need to unbolt a centre bearing mounted under the floor of the car to be able to move the propshaft far enough away. Do not pull the coupling out of the gearbox because this will sometimes allow the gearbox oil to drip out.

All that you have to do now is undo the ring of bolts which hold the differential into the casing (**fig 3**). Position yourself under the car so that you can pull the differential off its mounting studs and take the weight once it is free. But do not lie directly beneath it in case you find the weight too heavy. It is well worth having a helper on hand at this stage in case you get into difficulties.

The differential unit is quite heavy so it is important that you do not just pull it out of the axle and hope for the best. Use your hands to support the main weight at the back of the unit when it gets free of the casing. Bring the differential out from under the car so that you can push out the halfshaft stub and examine the axle as described below.

When you have got the differential out from under the

1. Scribing a line across the propshaft flanges

2. Unbolt the propshaft and swing it away from differential

3. Undo the ring of bolts or nuts at the front of the casing

4. With the halfshafts out the differential can be removed

car, check for any metal fragments caught in the works and make sure that it operates smoothly, then slide the new halfshaft into place in the side gear of the differential. The shaft should be an interference fit in the gear — it should need light pressure to slide in. When the shaft is in place, make sure that there is no movement between the splines on the shaft and the side gear. Any movement at all will mean that a shock loading will be imposed on the whole of the transmission and on the halfshafts especially when the clutch is let in and over bumps. A shock loading is usually what breaks the halfshaft in the first place.

If you can feel any wear or movement between the halfshaft and the side gear, first examine the splines on the shaft. Check that they are not showing any wear which will show up as either an obvious twist in the straight line of the spline or a waisted effect when looked at from the side.

When in doubt, try another halfshaft in the side gear. If there is still any movement you should consider buying a complete secondhand axle and fitting the whole thing to make sure that nothing is likely to cause another break. Do not risk using a loose halfshaft: it could break in a matter of days.

TIP

Fall out

If you think that there could be some metal fragments from the shaft in the axle tube, flush the tube with some paraffin from the wheel end so that the fragments end up in the bottom of the case. They can then be flushed out of the drain hole or be removed with a magnet.